ISO 9001:2000 Quality Management System Design

For a listing of recent titles in the *Artech House Professional Development Library,* turn to the back of this book.

ISO 9001:2000 Quality Management System Design

Jay Schlickman

Artech House
Boston • London
www.artechhouse.com

Library of Congress Cataloging-in-Publication Data

Schlickman, Jay J., 1934–
ISO 9001:2000 quality management system design
p. cm.–(Artech House technology management and professional development library)
Includes bibliographical references and index.
ISBN 1-58053-526-7 (alk. paper)
1. ISO 9000 Series Standards. I. Title. II. Series.
TS156.6 .S35 2003
658.5′62–dc21

2002038271

British Library Cataloguing in Publication Data

Schlickman, Jay
ISO 9001:2000 quality management system design—
(Artech House technology management and professional development library)
1. Quality control 2. ISO 9001 Standard
I. Title
658.5′62

ISBN 1-58053-526-7

Cover design by Gary Ragaglia

685 Canton Street
Norwood, MA 02062

International Standard Book Number: 1-58053-526-7
Library of Congress Catalog Card Number: 2002038271

10 9 8 7 6 5 4 3 2 1

This book is dedicated to my wife, Judith,
for forty-five years of total quality marriage and still going strong

Contents

Preface

The 20th century will be remembered as the Century of Productivity, whereas the 21st century will come to be known as the Century of Quality. So predicts Dr. Joseph M. Juran, father of the quality movement.

—Joseph A. DeFeo, "The Future Impact of Quality" Quality Engineering, Marcel Dekker, Volume 13, Number 3, 2001

Why We Are Replacing 13 Million Firestone Tires: Ford Motor Company is replacing all Firestone Wilderness AT tires on any Ford Motor Company vehicle. This action is a precautionary measure. Our analysis of real-world data, information from the federal government and lab testing indicate that some of the Firestone Wilderness AT tires not covered by last year's recall could, at some time in the future, experience increased failure rates.

— Ford Motor Company Advertisement, Friday, May 24, 2001, The Boston Globe, p. A27

Quality in a World of Globalization Without question, product quality is needed now as never before. Poor quality, especially in a world of globalization, equates to costs of nonconformance in the area of billions of dollars and, most importantly, oftentimes costs human life.

The pursuit of product quality requires that an organization create a quality framework. The ISO 9001:2000 quality management system (QMS) is an internationally established quality framework. This book is designed to assist an organization to structure an ISO 9001:2000 QMS on some well-established documentation and implementation concepts that have been proven to be effective based on ten years of intensive consulting and auditing experiences with 106 ISO 9000–certified organizations. The intent of this exercise is to provide the reader with a reasonable probability of maximized organizational productivity when the ISO 9001:2000 system is implemented.

The book's objective is to establish an engineering design approach to create a compliant ISO 9001:2000 QMS. Our design rules are constructed to effectively minimize documentation in a way that still increases implementation usage and fosters a dynamic demonstration of continual improvement.

An effectively designed QMS should do the following:

- Unify the organization's economic needs with its quality requirements;
- Optimize the flow of information to a wide range of users;
- Maintain full compliance with the ISO 9001:2000 International Standard (Standard);
- Provide a dynamic presentation of the organization's drive towards a meaningful ISO 9000 QMS;
- Propose a resolution as to just what a quality manual should contain and thereby provide a basis for a less diverse set of practitioner interpretations.

The book's approach is based primarily upon an interpretation of the requirements stated in the Standard and its associated guidelines. The directives are encased within the context of 39 years of experience in the management of high-tech research, engineering, marketing and sales, quality, manufacturing, and service organizations.

Although the design rules are generic, the text covers 1994–2000 upgrades in detail (the cut-and-paste and fill-in techniques) because over 400,000 1994-certified sites require upgrades prior to December 15, 2003. As a result, the upgrade requirement was used as the basis for the first case study. The second case study is designed for someone who needs to create a QMS from the ground floor using the book's design rules. The second case study is much shorter, as the first case study ends up with a complete quality manual that, in tone and structure, is similar to a first created quality manual.

Origins The ISO 9000 schema has matured to the point that it contains its own scholarship, mythology, and sibling conflicts. The program has transitioned to one of big business, complete with a plethora of international accreditation boards, registrars, trainers, and consultants under contract to thousands of global organizations. We have termed this group of entrepreneurs the ISO 9000 practitioners [1].

Every week, the ISO schema becomes more entrenched into the fabric of business and society (e.g., the certification of Nasdaq's computer and network operations, facilities and technical services to ISO 9001; the development of FS

9000 for the financial sector; and the United States Army's planned adoption of ISO 14001 by 2005 [2]).

The ISO 9000 practitioners work within an exciting and dynamic environment that now fosters a myriad pattern of standards and interpretations of those standards. It is this book's intention to make a significant contribution to the clarification of this broad range of perspectives—both for those who wish to create an effective QMS and for those who audit those systems.

It is my privilege, as an independent subcontractor, to work with this group of remarkable talents on both sides of the ISO 9000 street. This situation has afforded me the opportunity to serve as a consultant and both assessor and auditee within the ISO 9000 certification process. Hopefully, this has also provided me a more balanced view in my role as provocateur. My ISO 9000 experience with over 100 organizations has been extremely positive, and it is my wish to share this unique opportunity with the entire ISO 9000 community.

It is this book's contention that a successful implementation of ISO 9001:2000 in any type of organization is the result of a fully compliant and strategically driven QMS. The design platform described in the book consists of a set of design tools that can create a fully compliant QMS whose fabric is an organization's strategic business declaration.

In most cases, ISO 9000 QMSs are difficult to document, implement, and maintain. The greatest difficulty lies in the demonstration of continual improvement. The lessons learned during my experiences with over 100 systems should not be lost but should be documented for others to evaluate and utilize to create their own effective ISO 9000 QMS. The effort required to create a QMS that conforms to the 2000 revision is no less and perhaps a bit greater than a QMS that conformed to the 1994 version. However, the versions are decidedly different in structure and tone. We hope that this book will clarify the differences for the certified-experienced readers and establish a clear structural context for those readers in the midst of their first certification.

This book will prove useful to those organizations that have already created a QMS but would like to bring their efforts to a new level of effectiveness.

The single most difficult aspect in the creation of an effective QMS is the need to create documentation that addresses a broad audience. It is also the most difficult aspect of this design approach, and we have worked diligently to illustrate how a QMS can be designed to provide the required information for all system users.

Although the book has been written at a technical level designed to reveal the operational beauty and power of the Standard, the conceptual nature of the Standard is not easily envisioned because of its hierarchal nature and descriptive style. We have worked very diligently to clarify and to offer alternative ways to address such issues.

Specifically, the text has been written for a diverse audience comprising the following:

- Executives who wish to understand what an effective QMS looks like and want to ensure that the system is economically feasible and in concert with the organization's strategic goals;
- Members of steering committees, stewards, process champions, and ISO 9000 management representatives who must decide on the scope and design detail of the QMS configuration and who must ensure that the system is effectively implemented;
- Operational and audit team members who need to understand how to write an effective set of ISO 9000 documents and how to make sure that the system is measured effectively and contains a dynamic corrective and preventive action process;
- ISO 9000 practitioners who are interested in the study of self-consistent QMS configurations and what it is like to work on the other side of the table;
- Training course suppliers who can use the book as either a research source or as the day-to-day text.

Part Content This book establishes a set of design rules for effective QMS creation. In particular, the need for full compliance to each requirement (written as SHALL) of the Standard is addressed in detail. For completeness, several other system design configurations and strategies are also addressed, though in less detail. The overall structure of the book follows a hierarchal flow that first considers the total QMS design issue and then deals separately with the design of the quality manual, standard operating processes and procedures, work instructions, forms, and records, as well as a number of important supplemental design topics.

Part I establishes the basis for QMS design. It is imperative that the QMS be transparent to the overall strategic goals and objectives of the organization. To formalize this concept, this section deals with several possible choices upon which to base an integrated strategic and quality-based QMS design. The ISO 9001:2000 International Standard is chosen for further exposition because of its inherent international and national certification advantage. The fundamentals of ISO 9001:2000 QMS design are then discussed in detail (e.g., the three pillars of documentation, implementation, and demonstration of effectiveness that support QMS operational integrity; the QMS process model; continual/continuous improvement cycles; and mandatory documentation requirements).

Part II deals with QMS documentation design and establishes a four-tier documentation hierarchy as the basis for an effectively documented QMS. The critical role of the quality manual as a key driver to overall QMS effectiveness is discussed in detail. Then, the lower tier documentation (i.e., processes, procedures, forms, records, and other mandatory documents) is addressed in terms of optimum documentation structure and their specific roles in the QMS hierarchy.

Part III deals with QMS implementation and discusses organizational issues in regard to leadership, QMS planning, documentation implementation, and the impact of carefully planned internal audits.

Part IV describes the key change in philosophy from the previous ISO 9001 version, (i.e., the organization must now continually improve QMS effectiveness and accomplish this task via quantitative analysis of QMS performance). The critical area of quality objective design is then discussed in some detail in regard to formulation, implementation, and analysis.

Part V discusses QMS styles. The topics of inherent, broad readership requirements; the negative impact of a paraphrased manual; publication media choices, and effective writing styles are addressed to illustrate their impact on QMS effectiveness.

Part VI blends all of the tools together and summarizes their use in the creation of a fully compliant and strategically business-oriented QMS. This set of tools is deployed in the two case studies described in Part VII.

Part VII addresses the fact that there are over 400,000 1994 manuals that will need to be upgraded to the new Standard. Many thousands more will need to create their first manual in conformance with the Standard. As a result, we have created two case studies:

The first case study describes the upgrade and recertification of the Growth Corporation from ISO 9001:1994 to ISO 9001:2000. The exercise is based on a wholly fictitious (although you may spot yourself) but completely formed high-tech organization that utilizes this book's set of design tools. The corporation chooses a cut-and-paste and fill-in approach to electronically cut up the old manual:1994 into the new manual:2000. The result is a stand-alone form of quality manual in which the sections directly form a complete and compliant manual:2000 contained within this book.[1] Join the group and see how the Growth Corporation uses the cut-and-paste and fill-in method to upgrade

---

1. The choice of configuration is not meant to imply a so-called best approach. It simply represents the most common form of quality policy manual that I have found in working with over 142 manuals. The tier II, III, and IV documentation described is also based on the most common forms of processes, procedures, and forms that I have observed. I occasionally still come across integrated manuals. They are a problem for some third-party assessors because of their uniqueness, but that is the assessor's problem, not the supplier's.

their quality manual to the Standard. Of course there is a very wise consultant on board.

In case study #2, a friend of Growth needs to create their first QMS based on the Standard, and has come to Growth for advice. Growth's vice president of quality assurance comes to the rescue and offers a plan that has been used to achieve Growth's 1994 certification and an ISO 9001:2000 upgrade certification using the design tools presented in this book. The same wily consultant helps out.

Several appendixes are also used to present more detail with regard to tool application. Adherence to the proposed design rules will create a documented, implemented, and systems-effective QMS that is fully compliant with the Standard, and makes a powerful statement about the organization's technical competence, commitment to quality, and enterprise uniqueness.

Endnotes

[1] By the end of 1999, 150 countries had adopted ISO 9000. The estimated "ISO 9000 third-party registration industry has had a combined economic impact of $4 billion." This data was reported in *Quality Systems Update*, McGraw-Hill Companies, Vol. 10, No. 7, Fairfax, VA, July 2000. A later report by Quality Systems Update (*QSU Publishing Company*, Vol. 12, No. 7, July 2002, p. 1) indicates that the worldwide total of ISO 9000 registration certificates now stands at 510,616 in 161 countries.

[2] Reported in *Quality Systems Update*, McGraw-Hill Companies, Fairfax, VA, June 2001 and March 2001 publications; and in *The Environmental Management Report* by the same company in the June 2001 publication. Please note that the term certification is used to denote the receipt of a certificate from an ISO 9000 registrar. The registrar then places the site on their list of certified organizations and in this manner the organization is registered.

Acknowledgments

Many organizations and individuals have contributed to the creation of this book—and with the fear that all writers have in missing someone—I wish to thank the following groups who have had an inordinate positive effect upon my ISO 9000 perspective.

Accreditation and approval boards: The ASQ/ANSI Registrar Accreditation Board (RAB); Dutch Accreditation Board (RvA); and the Automotive Industry Action Group (AIAG).

Registrars: Bureau Veritas Quality International (North America), Inc., Jamestown, NY; Scott Quality Systems Registrars, Inc., Wellesley, MA; TUV America, Danvers, MA; and Intertek Technical Services (ITS), Boxborough, MA.

Consultant organizations/trainers: Information Mapping Incorporated (IMI), Waltham, MA, and its International Quality Systems (IQS) division; POWER Inc., Salem, MA; Management Software International, Inc., Stoneham, MA; and Corporate Development Services (CDS), Lynnfield, MA.

Direct clients: SMT East, now USANE, Middleborogh, MA; Dome Imaging Systems Incorporated, Waltham, MA; and ACS Technologies, San Diego, CA.

Artech House peer review: Artech House reviewer inputs were abundant and insightful, and many were incorporated into this book. However, any weakness in the text is purely my doing.

Special notice: I would also like to specifically thank the following colleagues for their many contributions, both knowingly and unknowingly, to this book (in no particular order). Dan Morgan, James R. DiNitto, Gary Deines, Hal Greenberg, Jerry Paradis, Steve Gaudreau, Frank Uttaro, Steve Zis, Bill Poliseo, Karl Titus, Stephen S. Keneally, Warren Riddle, Dr. Anthony F. Costonis, Ali Dincmen, Joe McCasland, Karen Snyder, John Bader, Mike Hayes, Robert J. Judge, Cas Makowski, Janet S. Cogdill, Bruce Mader, and Don Griffin.

PART

I

QMS Design Fundamentals

Strategic quality goals are established at the highest company levels and are a part of the companies' business plans. This concept of strategic quality goals is a logical result of the movement to give quality the top priority among the companies' goals.
—J. M. Juran, Juran on Quality by Design, The Free Press, 1992.

You shouldn't have a long-term strategy anymore, because you are going to be confined, and you won't be able to move fast enough.
—Orbit Gadiesh, Chairman, Bain & Co., "The Mind of the C.E.O, " Business Week, February 5, 2001, p. 108.

In the end, a vision without the ability to execute is probably a hallucination.
—Stephen M. Case, Chairman, AOL Time Warner, "The Mind of the C.E.O," Business Week, February 5, 2001, p. 107.

CHAPTER

1

Contents

QMS Foundations

1.1 The Relevance of Standards

Are management standards still relevant in a world of accelerated technology and rampant globalization? This issue is pertinent to any discussion of standards because standards are most useful when applied in a stable and predictable environment. If we operate under conditions of crisis and chaos we must use management techniques designed to handle large fluctuations. In the end, however, we still need a standard that defines our baseline so that measurements of our progress, or lack of progress, are meaningful.

To establish meaningful standards requires that there are universal organizational fundamentals. Such fundamentals must be constant, although the paradigms may shift (i.e., the way we model and apply the fundamentals varies with the most accepted global norms and mores). However, no matter what the paradigm shift involves, those who sell a product at a loss of one cent per piece will never make up the loss in volume. Those who do not know what their customer really needs will still lose to someone else who does. Those who do not cost-reduce their products continuously will eventually lose their market dominance. Those who do not periodically offer more performance for the same price will lose their competitive edge. And those who do not nurture their suppliers could lose a month's shipments waiting for a product from a vendor who went bankrupt because the vendor priced the product at a loss to win your contract.

Thus, the development and application of standards to enhance organizational development remains relevant in spite of the overwhelming, constantly changing twenty-first century explosion in technology and globalization. In fact, international and national standards are now in use in over 160 countries to form the foundation for effective quality management systems. The number of Annual Quality Awards now lists at least 119 programs worldwide. In the United States, statewide programs are underway in at least 41 states [1].

This book intends to present a systematic, engineering approach for the creation of effective QMSs. However, the framework for such systems requires knowledge of process-oriented structures. For this purpose, the following section discusses the concept of core competencies.

1.2 Core Competencies

The QMS requirements are superimposed upon the overall operational structure of the organization. You do not design the organization to follow a standard. Standards are used to enhance the effectiveness of the operating system. The operating system is designed to meet the needs of customers as dictated by the organization's market imperatives. The QMS is most effective when it is transparent to the overall strategic goals and objectives of the organization.

The strategic goals and objectives of the organization are embedded within the organization's processes or core competencies (i.e., the overall operational structure of the organization is in the form of core competencies) [2]. Each core competency is characterized by a process that must link seamlessly into the next core competency to produce an effective overall QMS. The model of a typical QMS is illustrated in Figure 1.1.

As indicated in Figure 1.1, the essential feature of the QMS is the conversion of customer requirements, as defined in a mutually agreed-to specification, into a product or service that satisfies the customer's applications. The critical supplemental feature of the QMS is the ability of the organization to

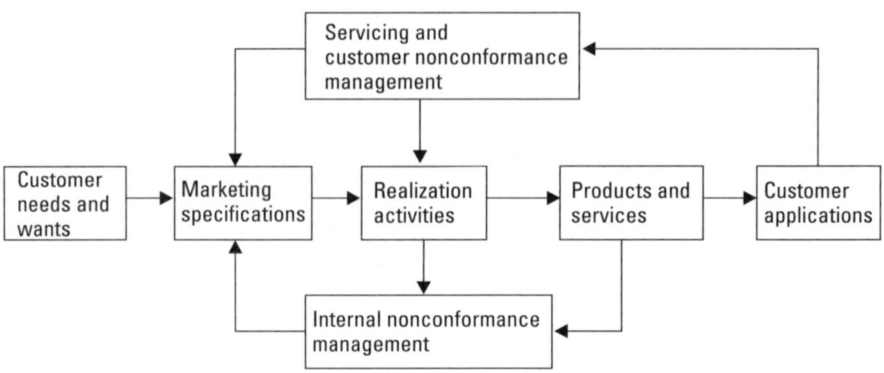

Figure 1.1 Functional model of a typical QMS.

measure and correct both its internal nonconformities that result from its realization activities and its external nonconformities that result from customer usage. The feedback loop entitled "internal nonconformance management" represents the internal nonconformities, and the feedback loop entitled "servicing and customer nonconformance management" represents the external nonconformities.

Both a core competency and a process transform inputs into usable outputs and are thus equivalent functional terms. However, the term process is more commonly used operationally and is more readily understood when the term *subprocess* is used.

1.2.1 Core Processes

In the development of an effective QMS, it is critical that all of the organization's core competencies (processes) are defined so that the overall management process is without gaps.

The interrelationships of the core processes form a spider web, and voids in the web are places where productivity and profits usually fall through. Personnel instinctively understand the workings of their own turf. The real problems arise when we seek to integrate turf-to-turf activities. A missed web ultimately results in a turf-to-turf conflict.

Figure 1.2 is an example of a typical set of enterprise core competencies that require a process document. A process document can be defined as a time-based description of the process that can be expressed in a flow chart or discussed in tabular form or in the form of a procedure.

Figure 1.2 Typical core competencies (processes and subprocesses).

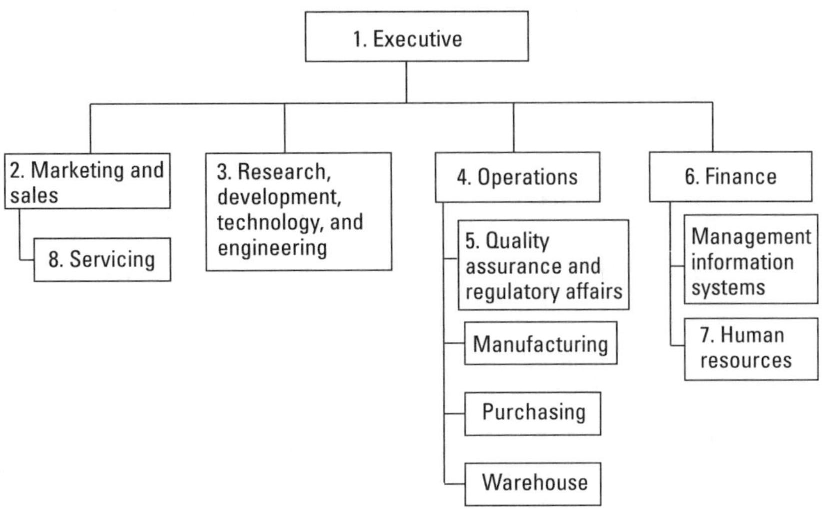

In Figure 1.2, there are eight core competencies defined. Core competency number 4 (operations) contains not only an additional core competency—quality assurance and regulatory affairs (QA&RA)—but also contains a number of subprocesses (e.g., manufacturing). As a result, the operations process charting would consist of an overall process that links up with the subprocesses. In this manner, all of the core competencies can be captured to form a complete QMS process.

The exact choice of core competencies—the resultant processes and subprocesses—is somewhat subjective and is a function of the economic impact of the function on the total organizational effectiveness.

For example, the management information systems (MIS) block under finance could just as well be placed under manufacturing, as it represents any number of computer systems that are used to analyze and control the enterprise's productivity and profitability.

1.2.2 Strategy To Transform Documentation into an Operational System

Once the organization's core competencies have been defined, it is necessary to select a strategy by which the now documented processes can be activated to form an effective QMS. Table 1.1 illustrates a typical strategy that systematically transforms the documented processes into an operational reality.

In this strategy, we are to create an effective QMS hierarchal structure that consists of the following:

- Documentation that accurately describes the organization's core competencies and provides the necessary policies, processes, procedures, forms, and records to support the organization's QMS;

Table 1.1
Typical Strategy To Achieve an Effective QMS

Levels	Create an Effective QMS Structure	Develop Employee Capability	Evolve the Effective QMS
1	An Effective QMS		
2	Demonstration of process effectiveness	Knowledge of QMS objectives, metrics, and targets (goals)	Quantitative management decision making
3	Implementation of core processes	Agreement on processes via ownership	Team-based management techniques
4	Process documentation of core competencies	Participation in the documentation process	Develop common QMS language

- Implementation based on the operational use of the documents on a daily basis;
- Demonstration of effectiveness based on the monitoring, measuring, and analyzing of operational data and the corresponding corrective and preventive action programs.

The activities of documentation, implementation, and demonstration of effectiveness form the three pillars upon which rest the quality management system's operational integrity (see Figure 1.3).

In parallel with QMS structure, employees develop knowledge of the organization's goals and objectives and the organization develops a common management language that results in a quantitative management style where decisions are based primarily on analyzed data. The content of this book develops a systematic approach to the application of this quantitative QMS strategy.

1.3 Selection of a QMS Baseline

A number of quality management baselines exist that can result in an effective QMS. They consist primarily of custom designed total quality management (TQM) programs, and programs built upon a nationally or internationally recognized standard. Figure 1.4 illustrates three specific types of systems for evaluation: a typical TQM example, a system based on ISO 9001:2000, and a system based on the Baldrige standard [3].

We note that all three systems can be designed to encompass all of the organization's core competencies. In addition, all three can employ action teams to measure cost of quality and to provide top management with a

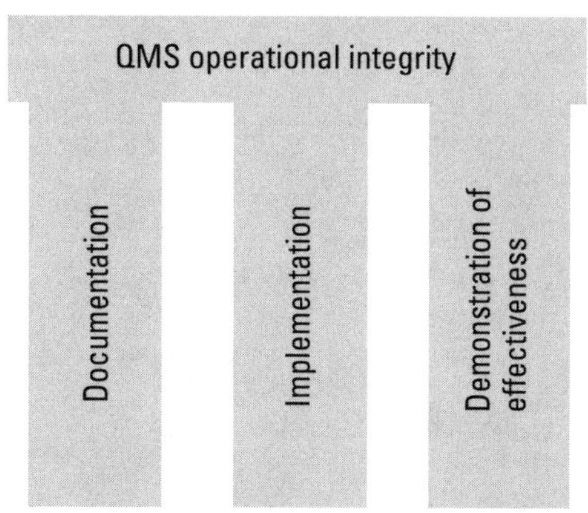

Figure 1.3 The three pillars of a QMS.

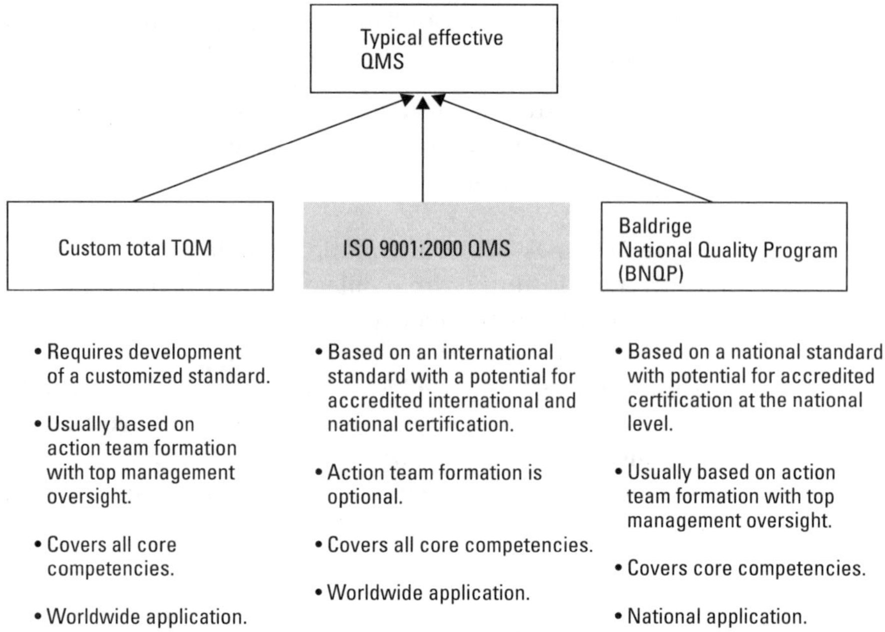

Figure 1.4 Potential effective QMS baselines.

corrective and preventive action protocol. The key difference between the three lies in the ability of the ISO 9001:2000 approach to attain certification, either nationally, internationally, or both.

Although the Baldrige award is a nationally recognized certification, of the three concepts, only the ISO 9001:2000 QMS provides for both accredited international and national recognition. In addition, the 1994 version is used as the foundation for a number of sector-specific applications that include the automotive, medical, aerospace, and telecommunications industries [4]. The adoption of the 2000 version as the foundation for the sector-specific standards is already underway.

Endnotes

[1] Johnson, Corinne N., "Annual Quality Awards Listing," *Quality Progress,* August 2001, p. 62, at http://www.asq.org.

[2] The term *core competency* is believed to have been coined by C. K. Prahalad, a University of Michigan Business School professor (see Stewart, Thomas A., "The World According to C.K. Prahalad," *Business 2.0,* January 2002, p. 92). It has been discussed in regard to ISO 9001:2000 by Jeanne Ketola and Kathy Roberts, "Demystifying ISO 9001:2000," *Quality Progress,* September 2001, p. 65.

[3] ISO 9001:2000, American National Standard, Quality management systems – Requirements, American Society for Quality, Milwaukee, Wisconsin, December 13, 2000; and Baldrige National Quality Program 2001, Criteria for Performance Excellence, NIST, Gaithersburg, MD. The Spring/Summer, 2001, ASQ Quality Press Publication Catalog lists 17 selections for TQM, at (http://qualitypress.asq.org).

[4] At this writing, this policy to use the latest ISO 9000 Standard as the basis for sector-specific standards will be followed for the medical device industry (i.e., ISO 13485) and will most likely be followed for all other sector-specific standards.

CHAPTER

2

Contents

The ISO 9001:2000 QMS

2.1 The ISO 9000 QMS Design Context

The process used to create an effective QMS based on the ISO 9001:2000 International Standard extends directly to the creation of any QMS based on a standard.

By a standard, we mean a document published by either a national or international organization that has achieved a relatively high level of industry recognition and credibility in its specific area of expertise. There are of course ad hoc standards that are created and distributed within specific technical fields. Such ad hoc standards are extremely useful but are generally not recognized at so high a national or international level.

Examples of widely recognized national and international standards include QS-9000 for the automotive industry; AS9100 and IAQS 9100:2000 for aerospace; the Baldrige National Quality Program for total quality management; TL 9000 for telecommunications; and the FDA/CGMP 820, EN46001, and ISO 13485 standards for medical devices.

In many cases, a specific standard is complemented by a series of additional mandatory standards. For medical companies that wish to deliver product into countries that require a product certification (CE mark), it is necessary to comply with the Medical Device Directive 93/43/EEC. Health Canada provides its own Medical Devices Regulations that require specific licensing. In addition, the ISO 14000 standard is used for environmental management systems.

QMS mastery is a journey not a destination. There are literally thousands of standards and supplemental guidelines in use throughout the world. However, no matter how complex the set of standards, the underlying process to create an effective QMS is the same. The mastery of this process is no different than the mastery of any technical regimen.

In our text, we focus this optimization process on the international standard, ISO 9001:2000 Quality Management System: Requirements [1]. Throughout the text, the term Standard (capitalized) is used to denote the ISO 9001:2000 International Standard.

2.2 Effective QMS Processes

The impact of ISO 9000 certification on performance is a popular topic for speculation. However, rigorous evidence of performance improvement and cost reduction has begun to appear in the literature [2].

The process to produce an effective QMS requires the following:

- The analysis of the tandard's requirements—these are stated in terms of SHALLS;
- The introduction of an interpretive scheme based on the author's experience and technical background;
- The top management decision on the total effort to be expended to produce the QMS (i.e., the degree of responsiveness);
- The integration of business strategy with strategic quality management goals;
- The clear presentation of the strategic organizational policies documented in a quality manual (manual);
- The aggressive implementation of the designed QMS;
- The demonstration that the QMS is effective through the analysis of data that tracks QMS performance against quality objectives.

The integration of business objectives with quality and customer satisfaction metrics—as the most effective way to evaluate corporate performance—is exemplified by the work of Robert S. Kaplan and David P. Norton with their publication in 1992 of *The Balanced Scorecard—Measures that Drive Performance*. Since then, it has been estimated by Bain & Co. that about half of *Fortune* 1,000 companies in North America use the Balanced Scorecard in their strategic analysis. Most importantly, the authors offer data that indicates that the technique produces positive results [3].

Practical Considerations In practice, ISO 9000 systems exist somewhere between the two limits of either a fully responsive QMS based on clearly defined and stated organizational policies or a QMS based on policies formed from just a repetition of the Standard's phrases. In my experience, the primary reason that fully responsive QMS structures are hard to find is that the documentation teams are unaware that there is a systematic design approach upon which to base their efforts. Once the teams are made aware of such an approach, their ability to optimize the flow of information throughout the QMS significantly improves. The response time in resolving organizational issues decreases and the overall gain in productivity improves via an enhanced knowledge by every employee on just what the organization's objectives are [4].

As a result, our goal is to present a set of QMS design rules that we believe can produce a fully responsive QMS that is both in compliance with the Standard and an effective strategic declaration of the organization's business objectives.

We firmly believe that the intrinsic value of the Standard is its bottom-line focus on productivity and thus profitability—regardless of how the supplier wishes to state such objectives (e.g., lowered customer complaints, increased return on investment, lowered rejects, increased repeat purchase orders, and lowered product-return rates).

The Standard—through its inherent continuous/continual improvement paradigm, stress on customer satisfaction, heightened awareness of a lowered cost of quality, transparent business/quality objectives, and explicit calls for process/procedural analysis—offers the supplier a unique opportunity to improve its competitive advantage.

Specifically, the Standard has integrated the following eight quality management principles into its requirements [5]:

1. Customer focus;
2. Leadership;
3. Involvement of people;
4. Process approach;
5. System approach to management;
6. Continual improvement;
7. Factual approach to decision making;
8 Mutually beneficial supplier relationships.

As a result, only a fully responsive QMS will include the totality of the eight principles and offer the organization the maximum return against these principles. However, this potential for enhanced marketability, productivity, and profitability is dependent upon the supplier's desire to fully comply with the Standard, write the documented system in a user-friendly manner for a very wide range of readers, make a total management commitment to this effort, and establish a QMS that can be maintained in a cost effective manner.

The goal is to improve organizational effectiveness, not just get certified. Most importantly, a unified, strategic, business-and-quality policy signals to all employees that the main purpose of the ISO 9000 certification is to improve the effectiveness of the operation, not just achieve certification.

2.3 The ISO 9000 QMS Process Model

The manner in which the Standard achieves continual improvement is by means of its process orientation. The roots of this process are inextricably wound into the QMS definition.

2.3.1 Quality Management System Defined

The characteristics of a QMS in regard to quality include the following [6]:

1. The establishment of policy and objectives by an organization to manage resources;
2. The assignment of responsibilities and authority to personnel;
3. The development of a an organizational structure among the personnel.

2.3.2 Operational Model for ISO 9001:2000

Based on this definition, we can graphically demonstrate the functional relationships between the various parts of a QMS. This concept is shown in Figure 2.1.

The difference between Figure 2.1 and Figure 1.1 is that Figure 2.1 explicitly lists the appropriate section number for each activity. This means that Section 4.0 of the Standard is indicated where it is not included in the Standard's model. Figure 2.1 also indicates the benefits to the enterprise in terms of increased profitability, productivity, and product performance [7].

Figure 2.1 also integrates the three pillars of ISO 9000 (i.e., the documented system, its implementation, and its demonstration of effectiveness).

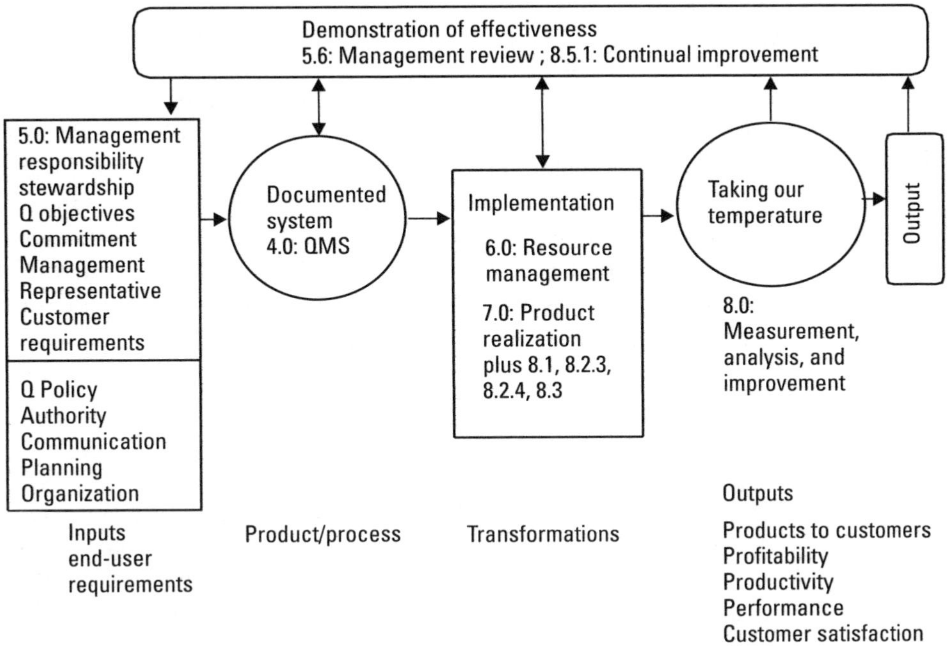

Figure 2.1 Operational model for the ISO 9001:2000 QMS.

This is not meant to imply that one model is better than the other. We do mean to clearly illustrate how our operational approach adheres in detail with the Standard's model. In fact, those who have either created the ISO 9001:2000 QMS already or are in the process will often format their process discussions in terms of 5.0 Management Responsibility; 6.0 Resource Management; 7.0 Product Realization; and 8.0 Measurement, Analysis, and Improvement rather than in terms of core competencies. Unfortunately, the 5.0, 6.0, 7.0, 8.0 approach can bypass the key process requirements of Section 4.0, Quality Management System.

Although I have found little difficulty with the use of the Standard's sections (instead of core competencies), the approach seems to need a more extensive, careful set of reference links to send the reader from one process to another as compared to core competencies that tend to automatically link functions. But this is really more style than substance.

Regardless of which model you choose, you will always have to integrate into the flow support functions such as management review, control of documents, control of records, control of monitoring and measuring devices, internal audit, and corrective and preventive action. Core competencies tend to highlight these support functions more—witness the missed Section 4.0 in the Standard's model.

We see in Figure 2.1 that the Standard has essentially defined a classic engineering feedback system complete with inputs, outputs, and feedback loops. The inputs of end-user requirements, quality objectives, and quality management protocols are framed by the documentation system and transformed by the implementation system to produce continuously improved processes and products. These lead to outputs that include enhanced products, productivity, profitability, performance, and customer satisfaction.

In summary, between the Standard's process model and our operational model—in concert with our plan-do-check/study-act models—it is possible to graphically display the most important aspects of the ISO 9001:2000 requirements designed to create continual improvement.

Endnotes

[1] References to ISO 9000 documentation are based on either the *Ninth Edition of the ISO Standards Compendium,* ISO 9000 Quality Management, International Organization for Standardization , Geneva, Switzerland, www.iso.ch; or the American National Standard series by the ANSI/ISO/ASQ. They are equivalent.

[2] See, for example, Romano, P., "ISO 9000: What Is Its Impact on Performance?" *QMJ,* Vol. 7, No. 3, 2000, ASQ.

[3] "On Balance," *CFO,* February 2001, p. 73; Lawton, Robin, "Balance Your Balanced Scorecard," *Quality Progress,* March 2002, p. 66, at http://www.ASQ.org.

[4] Personal observation: One of the key questions that I ask during surveillance audits is in regard to what has been the most dramatic impact of the QMS on an organization. Invariably the answer has to do with greatly improved overall communication in regard to problem solving. The second highest frequency response is a growing knowledge of organizational progress against goals.

[5] ISO 9000 Quality Management Principles at http://www.iso.ch/iso/en/iso9000-14000/iso9000/qmp.html.

[6] Re: ISO 9000:2000, Clause 3.2.3.

[7] Guidance on the Process Approach to Quality Management Systems at http://www.bsi.org.uk/iso-tc176-sc2.

CHAPTER

3

Contents

QMS Continual Improvement Framework

3.1 Continuous/Continual Improvement Is Inherent

We have established that the QMS should be a blend of business strategy and quality management (an integrated QMS)—in full conformance with the Standard. This section creates the implementation framework for our approach.

3.1.1 Continuous Versus Continual Improvement Concept

First it is necessary to understand the equivalency between continuous and continual improvement so that we can readily use the terms interchangeably.

The normative definition in ISO 9000:2000 for continual improvement states that it is a recurring activity to increase the ability to fulfill requirements (3.1.2) [1]. The definition notes that the process (3.4.1) of establishing objectives and finding opportunities for improvement is a continual process through the use of audit findings (3.9.5) and audit conclusions (3.9.6), analysis of data, management reviews (3.8.7), or other means and generally leads to corrective action (3.6.5) or preventive action (3.6.4). This is basically the plan-do-study-act scenario that was originally described in a less explicit manner by the American physicist Walter A. Shewhart in 1931 [2].

The equivalency of the two ideas can be readily shown by comparing this activity to a typical continuous improvement scenario used in TQM programs (see Table 3.1) [3].

3.1.2 Quality As a Philosophy [4]

As a result, the Shewhart cycle can be used as the basic tool for continuous/continual improvement and as the foundation for QMS implementation. However, before we can logically define a method for QMS implementation, we must first define what we mean by a QMS, and, in particular, what we mean by quality. A quality system must be designed to be measurable. In fact, the cost of poor quality can be staggering [5].

Table 3.1
Equivalency of Continuous and Continual Improvement

Shewhart Cycle	Continual Improvement Process As Specified in ISO 9000:2000	Continuous Improvement Process As Specified in a Typical TQM Program
Plan (initial)	Top management formulates a continual improvement process based on quality objectives and a search for opportunities for improvement	Top management formulates a continuous improvement process based on cost-of-nonconformance metrics and a search for opportunities for improvement (OFIs)
	The management review is used for decision making	The executive review committee is used for decision making
Do	Action items are assigned by top management to resolve problem areas	Action teams are assigned by the executive review committee to resolve problem areas to develop a project schedule, and to identify milestones for completion
Study	Audits findings and conclusions, as well as other databases, are used as the basis for analysis of data	Root causes are identified by means of interviews, internal audits, and supplier audits
		A system is developed to measure the results of a proposed corrective action
Act	Corrective or preventive actions are taken and presented at management review	A list is developed of possible corrective actions and one solution is selected
		A system is developed to implement the corrective action
		The proposed plan is implemented on a test basis and monitor the results to determine the proposed action's effectiveness
		Oversight is provided by the executive review committee
Plan (iterative)	The process is repeated—another problem area is selected for resolution	The process is repeated—a method is proposed to implement the corrective action companywide

3.1.3 Quality As a Scientific Measurement

The ISO 9000:2000 vocabulary specifies quality in operational terms. (Alert: the definitions are normative—they are part of the Standard, not just a guideline). The definition begins with the word *degree* [6].

Degree in the definition implies a scale associated with quality. Quality is not absolute but relative to what is acceptable from the receiver's (customer's) standpoint. We also learn from the definition that quality is based upon not only what the customer needs but also what the customer expects. This is what makes the fulfillment of quality so difficult—few of us really fully know what we need. Until we receive the result of the contract, we do not really know what our expectations are, even when there is a specification. Just try meeting someone's expectations in regard to paint and you will find that gloss and aesthetics are one big headache.

As a result, I consider quality an iterative process that depends upon specific measurements but that is always open to improvement. This is the important role of validation (i.e., a test program that includes the customer's participation whenever possible).

Thus, when we begin the quality process we mean that high quality is defined as our ability to meet customer requirements that have been specified quantitatively. For a service organization, it might mean 7-day, 24-hour ready availability or an effective triage to provide health management. For a manufacturing organization, it might mean on-time delivery, or user-friendly instrumentation. All of the requirements must be measurable and addressable in terms of metrics. Otherwise, you have an open-ended relationship, and nobody knows when the job is done and when it is time to get paid—a common problem in contracting for either a new sun deck for your house or a QMS [7]. Whatever the metrics are, they must be subject to analysis and continual improvement. Such metrics form the basis for enterprisewide quality objectives [8].

3.1.4 Continual Improvement Is Intrinsic Within the Standard

The ability to define a continually improving (C/I) QMS is inherent in the Standard, and the Standard's process orientation provides a method to drive the QMS at whatever rate makes sense for the organization.

3.1.5 Customer-Driven Orientation

The customer orientation of the Standard was introduced when we stated the eight quality management principles, the first of which is customer focus (where customer refers to interactions between both internal and external parties). For example, the requirements for a customer-driven program are fortified in a number of the Standard's clauses, including 5.1 Management commitment, 5.6.2 Review input, 5.6.3 Review output, and 6.1 Provision of resources. The essence of these clauses deal with communication in regard to meeting customer requirements, customer feedback, and the enhancement of customer satisfaction.

In this manner, we can demonstrate that the Standard provides us with the platform for a unified QMS because the Standard's orientation is thematically aimed at an effective customer relationship.

Next, we need to demonstrate how the continuous improvement cycle—desired by both ourselves and the customer—is intrinsic within the Standard.

3.1.6 Shewhart Cycle

We can demonstrate the inherent continuous/continual improvement properties of the Standard if we indicate the relationship between the five operational sections and their corresponding paragraphs of the Standard and the Shewhart cycle of plan-do-check-act as indicated in Figure 3.1.

In this diagram, we have placed each of the operational paragraphs in a related category of the Shewhart cycle [9]. The exact placement of the elements is subject to conjecture, but what is important here is that there is an approximate 1:1 correspondence with the paradigm [10].

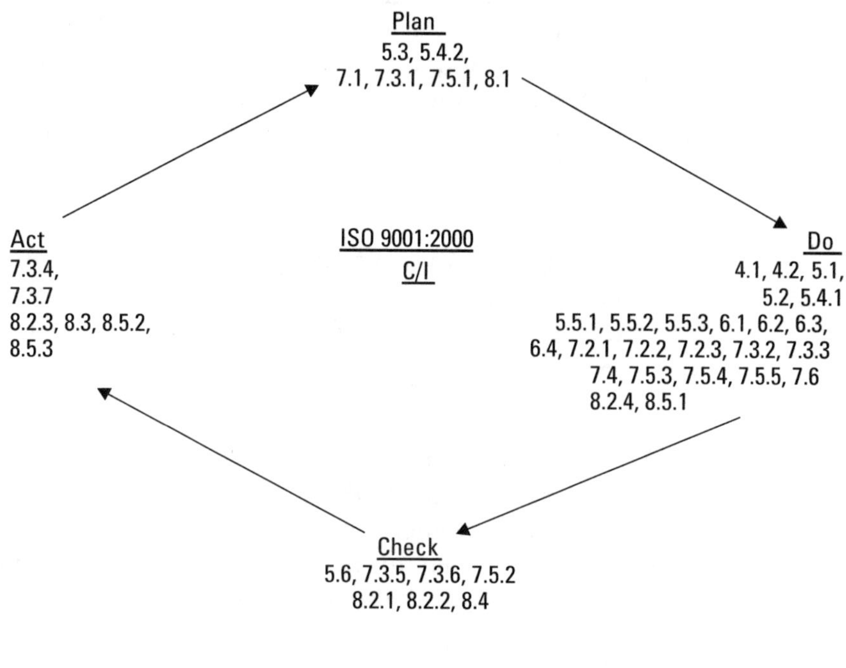

Figure 3.1 ISO 9001:2000 continuous/continual improvement cycle by paragraph.

Plan. The following Standard paragraphs provide the framework in which top management places its unified quality/business plans, marketing, and sales promotions and strategies, establishes performance metrics, and records progress against goals to measure the effectiveness of the QMS:

- 5.3 Quality policy;
- 5.4.2 Quality management system planning;
- 7.1 Planning of product realization;
- 7.3.1 Design and development planning;
- 7.5.1 Control of production and service provision;
- 8.1 General (measurement, analysis, and improvement).

Do. The following Standard paragraphs establish the implementation protocols within which we design, manufacture, and service products:

- 4.1 General requirements;
- 4.2.1 General (documentation requirements);
- 4.2.2 Quality manual;
- 4.2.3 Control of documents;
- 4.2.4 Control of records;
- 5.1 Management commitment;
- 5.2 Customer focus;
- 5.4.1 Quality objectives;
- 5.5.1 Responsibility and authority;
- 5.5.2 Management representative;
- 5.5.3 Internal communication;
- 6.1 Provision of resources;
- 6.2 Human resources;
- 6.3 Infrastructure;
- 6.4 Work environment;
- 7.2.1 Determination of requirements related to the product;

- 7.2.2 Review of requirements related to the product;
- 7.2.3 Customer communication;
- 7.3.2 Design and development inputs;
- 7.3.3 Design and development outputs;
- 7.4 Purchasing;
- 7.5.3 Identification and traceability;
- 7.5.4 Customer property;
- 7.5.5 Preservation of product;
- 7.6 Control of monitoring and measuring devices;
- 8.2.4 Monitoring and measurement of product;
- 8.5.1 Continual improvement.

Check. The following Standard paragraphs provide the mechanisms whereby we monitor our progress against quality goals so that the entire QMS can be analyzed and corrected to achieve continual improvement:

- 5.6 Management review;
- 7.3.5 Design and development verification;
- 7.3.6 Design and development validation;
- 7.5.2 Validation of processes for production and service provision;
- 8.2.1 Customer satisfaction;
- 8.2.2 Internal audit;
- 8.4 Analysis of data.

Act. The following Standard paragraphs establish the methods required to correct those areas that are out of conformance and to establish long-term preventive action programs:

- 7.3.4 Design and development review;
- 7.3.7 Control of design and development changes;
- 8.2.3 Monitoring and measurement of processes;
- 8.3 Control of nonconforming product;

- 8.5.2 Corrective action (w/response to customer complaints);
- 8.5.3 Preventive action.

There is operational power when all clauses are implemented. Thus, when all paragraphs of the Standard are implemented, the paradigm ensures that the system will be documented; that those documents will be used by the employees; and that there will be adequate measurements made to judge whether or not we have demonstrated effective performance against our business/quality objectives.

3.2 Continuous Improvement Cycle Within Elements

3.2.1 Other C/I

The continuous improvement cycle can also be demonstrated in specific sections of the Standard (e.g., Section 7.3: Design and Development, as shown in Figure 3.2).

3.2.2 Further Demonstration

We can also demonstrate that Section 7.5: Production and Service Provision (P&SP)—as illustrated in Figure 3.3—also contains a continuous improvement cycle.

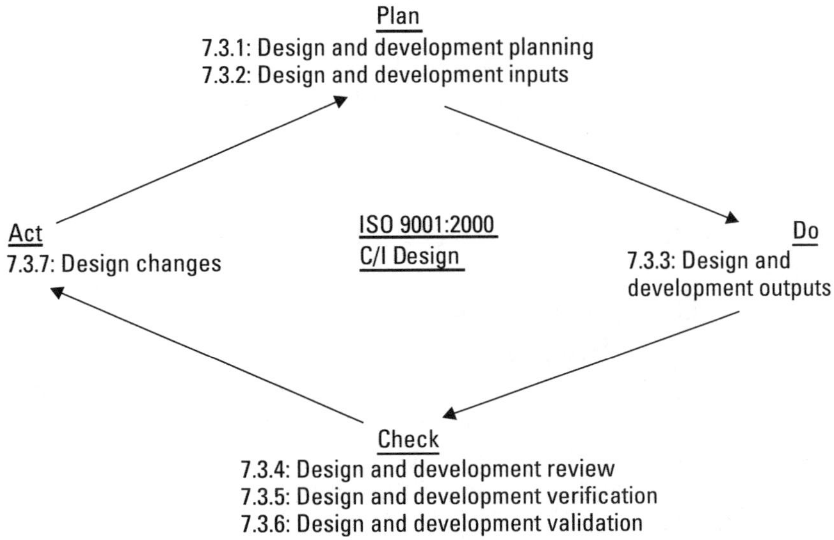

Figure 3.2 Section 7.3: Design and Development—continuous improvement cycle.

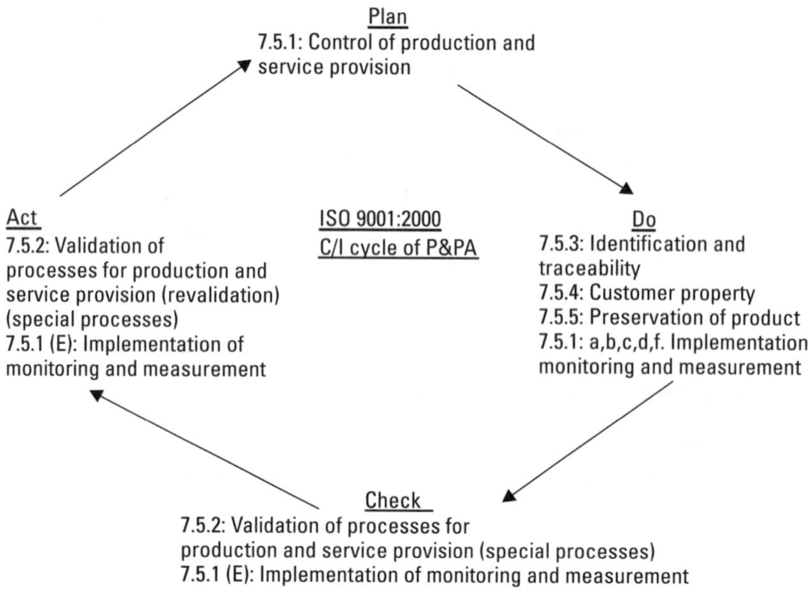

Figure 3.3 Section 7.5: Production and Service Provision.

3.2.3 Continuous Improvement Cycle

This phenomenon is a general trend throughout the Standard as demonstrated further in Table 3.2. This table is not meant to be inclusive, but illustrates the general trend. The interested reader, who delves deeply into the ISO well, will find even more clauses that fit the cycle.

3.2.4 Continual Improvement Imperative

We conclude that both a market orientation and the continuous improvement cycle is inherent within the Standard—whether you wish it or not—and as a result it is necessary to respond to every requirement to ensure that the Standard's continual improvement integrity is maintained.

3.3 Mandatory Documentation Requirements

The creation of a QMS—based on the Standard—requires a fully compliant documentation system (i.e., a QMS in which each SHALL of the Standard is clearly documented).

The desire to integrate business and quality objectives, so that they are transparent, is a repetitive theme throughout the Standard and its associated guidelines (see Figure 3.4) [11].

In the ISO 9000:2000 schema, the documents are intended for the following:

Table 3.2
Examples of Other Elements That Contain the Continuous Improvement Cycle

ISO 9001:2000 Element	Plan →	Do →	Check/Study →	Act →
7.4 Purchasing	Establish criteria to evaluate and select suppliers.	Select suppliers and prepare purchasing information.	Ensure adequacy of specified purchase requirements and maintain records. Implement the inspection or other activities.	Evaluate and reevaluate suppliers.
8.5.2 Corrective Action	Document procedure to define requirements.	Review nonconformities (including customer complaints); determine the causes.	Evaluate the need for action; determine action needed. Review corrective actions taken.	Implement actions needed and record the results of actions taken.
8.2.2 Internal Audit	Planned program at planned intervals and planned arrangements. Define audit criteria, scope, frequency, and methods. Create documented procedure.	Conduct internal audits based on status and importance of the processes and areas to be audited. Maintain records. Effectively implement and maintain QMS.	The management responsible for the area ensures that actions are taken without undue delay to eliminate detected nonconformities.	Follow-up activities to include the verification of actions taken and the reporting of verification results.

- ISO 9004:2000, entitled "Quality Management Systems—Guidelines for Performance Improvements" is to be used to design the QMS.
- ISO 9001:2000 (Standard), highlighted in the center, is to be used for all contractual agreements.
- ISO 9000:2000, entitled "Quality Management Systems—Fundamentals and Vocabulary" is to be used as part guideline and part standard because the terms and definitions given in the document apply to the Standard. This important point is often overlooked by practitioners.

3.3.1 Accreditation Impact on Guidelines

Contrary to common belief, guideline documents are sometimes specified by the Accreditation Boards—via the Registrars—as strict requirements for certification, either as a constraint on the Registrar or on the organization. Two typical examples are as follows:

Figure 3.4
A portion of the ISO 9000:2000 Standard's documentation.

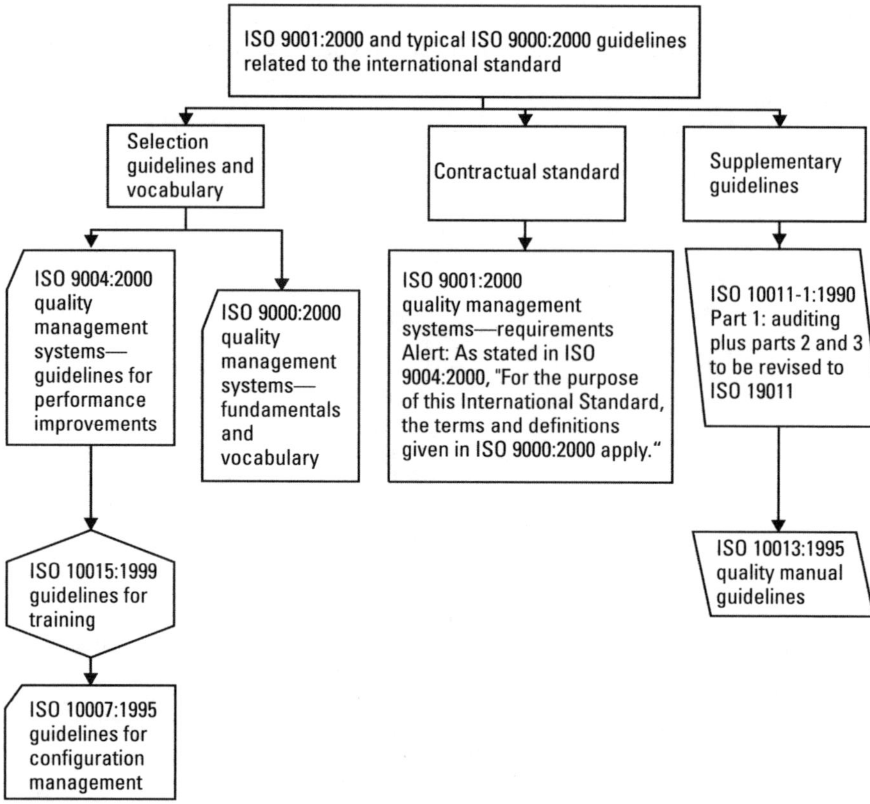

1. *ISO 19011: in process*—guidelines on quality and/or environmental management systems auditing;

2. *EN 45012: September 95*—general criteria for certification bodies operating quality system certification (e.g., Clause 18 requires the supplier to keep a record of all customer complaints and corrective actions taken in regard to such complaints). The European Normal (EN) series consists of many supplementary ISO documents.

The ISO 9000 family of documents focuses its guidance and requirements on satisfying the customer, and this motif is exemplified in the guidelines by stipulating that the organization's leadership should actually create a customer-oriented organization [12]. Thus, we have a clear indication of the concept of a unified business/quality imperative as a prime directive of the Standard's intent.

It is not by accident that the Standard stresses the unification of quality and strategic business objectives. The development of the eight quality management principles is a result of a concerted effort by the United States ISO TC 176 technical committee to create a unified set of principles based on research using quality-related documents from all over the world. The Standards analyzed include the Baldrige award.

Scott Madison Paton notes that, "From 1990 to 1999, the publicly traded recipients [of the Baldrige Award], as a group, outperformed the Standard & Poor's 500 by 4.2 to 1, achieving a 685.26% return compared to a 163.11% return for the S&P 500" [13]. Paton's analysis is another indication of the bottom-line focus of ISO 9000 [14].

The similarity to the Baldrige National Quality Program 2002 is striking, and this fact has been noted by several authors [15]. An interpretation is offered in Table 3.3 in which we compare the Baldrige sections with both the eight quality management principles and the pertinent sections of the Standard. As you can see, the correlation in theme and intent is obvious.

Table 3.3
Similarity of Baldrige Criteria for Performance Excellence Versus the ISO 9001:2000 Standard

Baldrige Section	ISO: Eight Quality Management Principles	ISO: Standard's Requirements by Section
1.0 Leadership	b) Leadership f) Continual Improvement	5.0 Management Responsibility
2.0 Strategic Planning	e) System Approach to Management	5.0 Management Responsibility 6.0 Resource Management
3.0 Customer and Market Focus	a) Customer Focus	5.0 Management Responsibility 7.0 Product Realization 8.0 Measurement, Analysis and Improvement
4.0 Information and Analysis	g) Factual Approach to Decision Making e) System Approach to Management	5.0 Management Responsibility 7.0 Product Realization 8.0 Measurement, Analysis and Improvement
5.0 Human-Resources Focus	c) Involvement of People	6.0 Resource Management
6.0 Process Management	d) Process Approach h) Mutually Beneficial Supplier Relationships	4.0 Quality Management System 7.0 Product Realization 8.0 Measurement, Analysis and Improvement
7.0 Business Results	g) Factual Approach to Decision Making	5.0 Management Responsibility 8.0 Measurement, Analysis and Improvement
Source: [16]	*Source:* [17]	*Source:* [18]

To fully appreciate the Standard's umbrella-documentation complexity, it is necessary to summarize all of the mandates so that a proper analysis can be achieved. The requirements are summarized in Table 3.4. The 1994 clauses are indicated in italics. I have found such cross-references to ISO 9001:1994 most useful in upgrading 1994 quality manuals to the 2000 version. (For an illustration of the concept of tiers, see Figure 4.1.)

Although there are a large number of different types of documents required by the Standard, our design approach offers a highly disciplined and logical approach to QMS documentation structure. For this purpose, a clearly defined and consistent taxonomy is required that is based upon long-established guidelines on how to propagate technical information effectively. All of the required documentation is readily incorporated into this versatile documentation structure. Accordingly, the taxonomy used in this book is defined in Table 3.5. The application of this structure is described in Part 2 [19].

It is important to note that records are filled-in and filed forms that can occur at any documentation level. They constitute a separate document category, and a separate set of control rules are required (refer to Par. 4.2.1(e) of the Standard). Several typical records are indicated to clarify this issue. The subject of records is covered more fully in Section 8.1.

3.3.2 QMS Design Methods To Be Presented

To accomplish this goal (i.e., to produce a fully responsive QMS in compliance with the Standard—that also integrates the organization's strategic and quality objectives—this book describes a series of ISO 9001:2000 QMS design rules that prescribe methods to enhance clarity, user friendliness, and compliance). Such methods include the following:

- The integration of business strategy with quality management;
- The use of the inherent continuous/continual improvement cycle;
- The need for stewardship;
- The development of effective QMS documentation structures;
- The avoidance of paraphrasing;
- The use of different documentation media;
- The development of prescriptive quality policy statements;
- The SHALL analysis method;
- The quality manual sequence methods;

Table 3.4
Summary of the ISO 9001:2000 Mandatory QMS Documentation Requirements

Tier Level	2000 Standard's Clause	Standard's Requirements Related to Documentation	X-ref to 1994 Standard
I	4.2.1 b)	A quality manual that contains a scope and justified exclusions.	4.2.2
	4.2.1 a); 5.3 5.4.1	Documented statements of a quality policy and quality objectives ... (separate or in the quality manual)	4.1.1 4.2.2
	4.1 a)	Identification of the processes needed by the QMS and their application throughout the organization ...(separate or in the quality manual)	4.2.1
	4.1 b)	Determination of the sequence and interaction of such processes ... (separate or in the quality manual)	
	4.2.2	A description of the interaction between the processes of the QMS ... (included in the quality manual).	
	5.6.1	A top-management review of the organization's QMS at planned intervals ... (tier I record)	4.1.3
II	7.1	The manner in which the organization plans the processes needed for product realization (quality or control plan)	4.2.3 4.10.1
	8.1	The manner in which the organization plans the monitoring, measurement, analysis, and improvement processes needed (quality or control plan)	4.10.1 4.20
	4.1	Identify the control of outsourced processes	4.6.1
	4.2.1 c)	Documented procedures required by the Standard (contained in either the quality manual or references to them)	4.2.2
	4.2.3	1. Control of documents procedure	4.5
	4.2.4	2. Control of records procedure	4.16
	8.2.2	3. Internal audit procedure	4.17
	8.3	4. Control of nonconforming product procedure	4.13
	8.5.2	5. Corrective action procedure	4.14.1 4.14.2
	8.5.3	6. Preventive action procedure	4.14.1 4.14.3
III	7.5.1 b)	Controlled work instructions, where applicable and necessary	4.9, 4.19 4.15.6
All	4.1	A documented QMS	4.2.1
	4.2.1 d)	Additional documents needed by the organization to ensure the effective planning, operation, and control of its processes (This is the "sleeper" requirement that drives the creation of a multitude of documents!)	4.2.2
	4.2.1 e)	Records required by the standard	4.2.2 4.16
II and IV	7.5.1 a)	Controlled information that describes the characteristics of product, where applicable	*4.9* 4.19, 4.15.6
	8.2.2	Planned intervals for the Internal audits	4.17

Table 3.5
Taxonomy Used in This Book (with Typical Types of Documents Noted)

Document Tier	Book's Terminology	Typical Content	Typical Record(s)
I	Quality manual (same definition as Standard)	Policies (i.e., response to SHALLS), process summaries and interactions, justification for exclusions, quality policy statement	Management reviews, management representative memo
II	Process documents (same definition as Standard)	Processes, standard operating procedures (SOPs), process flow charts, quality plans, control plans	Master document lists, device master record, device history record, design history files, technical files
III	Procedures (same definition as Standard)	Work instructions, work directions, test plans, calibration and maintenance plans	Quotes, sales orders, travelers and routers, inspection and test data, certificates of conformance and analysis
IV	Forms [see 4.2.1d) of Standard]	Templates and formats onto which drawings, blueprints, labels, test plans, and data are placed	CAD drawings, corrective and preventive action reports, job descriptions

- The possible quality manual configurations;
- The sector-specific requirements prescribed by ISO 9000 accreditation boards.

The reader, armed with this set of design rules, should be able to create an ISO 9001:2000 QMS that represents the true nature of the organization and supports its competitive advantage. In that regard, the first set of design rules is presented in Part 2.

Endnotes

[1] Guidance on the terminology used in ISO 9000:2000: http://www.iso.ch/iso/en/iso9000-14000/iso9000/2000rev8.html.

[2] The role of the Shewhart Control Chart and the development of the continuous improvement cycle is introduced in both Juran, J. M., *Juran on Quality by Design*, The Free Press, 1992, and Deming, W. Edwards, *Out of the Crisis*, Cambridge, MA: MIT Press, 1986. Dr. Deming is very clear on the fact that the Deming cycle is based on the original Shewhart cycle.

[3] Based on the work of Dr. Anthony F. Costonis, president and founder of Corporate Development Services, Inc., of Lynnfield, MA, at http://www.Corpdevelopment.com.

[4] See, for example, Hoyer, R.W., and Brooke B. Y. Hoyer, "What is Quality?" *Quality Progress,* at http://www.asq.org, July 2001, p. 53.

[5] Joann Muller summarizes the cost of poor quality for Ford in *Business Week,* June 25, 2001, p. 80. Ms. Muller notes that "Quality problems and related production delays cost the company more than $1 billion in lost profits last year alone." The Firestone tire fiasco "cost Ford about $500 million."

[6] ISO 9000:2000: Clause 3.1.1 Quality.

[7] Whenever I do a certification audit, I do not ask what the quality policy is. This would require rote memorization. Instead, I ask what quality means to the employee. The answers are enlightening (e.g., "Looks perfect, nice shape, good condition," "Do it right the first time," "Product that is workable," "The way it is supposed to be," "Get a repeat order," "Make the customer happy," "Something I would buy myself," and "Meet customer specs").

[8] The pervasiveness of measurement in the Standard has, rightfully, become a popular publication topic. See, for example: Olivier, Daniel P., and Paschal Dwane, "Using Measurement to Improve Quality," *Medical Device and Diagnostic Industry,* October 2001, p. 84, at http://devicelink.com/mddi.

[9] The integration of the Shewhart cycle by the ISO/TC 176 committee is discussed in, "Guidance on the Process Approach to Quality Management Systems," ISO/TC 176/SC 2/N544, December 2000, at http://www.iso.ch/ 9000e/2000rev_9e.htm.

[10] For a lucid discussion of the various applications of the Shewhart cycle refer to "What Deming Told the Japanese in 1950," *QMJ,* Fall 94, p. 9.

[11] The ISO 9000 Family: http://www.iso.ch/iso/en/iso9000-14000/iso9000/selection_use/iso9000family.html.

[12] Section 1.4 of ISO 9004:2000.

[13] Paton, Scott Madison, "Will the Baldrige Endure?" *Quality Digest,* May 2001, p. 4, at http://www.qualitydigest.com.

[14] See, for example, Hsu, Sangem "ISO 9004 Finally Measures up to Excellence," *Quality Systems Update,* February 2001, p. 15, and West, Jack, "ISO 9000:2000—Direction, Changes, and Opportunities," *European Quality Institute,* 2000, Section 8, http://www.iso9000directory.com.

[15] See, for example, Marash, Stanley A., "Fusion Management, Part One, Comparing Business Process Models," *Quality Digest,* June 2001, p. 22, *The Memory Jogger 9000/2000,* GOAL QPC, p. 157, at http://www.goalqpc.com, and Tonk, Hampton Scott, "Integrating ISO 9001:2000 and the Baldrige Criteria," *Quality Progress,* August 2000, p. 51.

[16] Baldrige National Quality Program 2002, "Criteria for Performance Excellence," NIST, Technology Administration, Department of Commerce, Gaithersburg, MD, at http://www.qualilty.nist.gov.

[17] American National Standard, Quality Management Systems—Fundamentals and Vocabulary, ISO 9000:2000, American Society for Quality, Milwaukee, WI, December 13, 2000, p. ix, at http://qualitypress.asq.org

[18] American National Standard, Quality Management Systems—Requirements, ISO 9001:2000, December 13, 2000.

[19] Our documentation taxonomy is based on the work of Robert E. Horn, founder of Information Mapping, Inc., Waltham, MA, at http://www. infomap.com. See, for example, "Mapping Hypertext," Lexington Institute, 1989.

PART

II

QMS Documentation Design

Taking on too much at once can sap any amount of energy and thwart the successful completion of any undertaking. Parceling tasks into manageable portions without losing sight of the overall design of the whole endeavor is one of the arts of leadership at all levels, from personal self-management to corporate, community, and political domains of action.

—Thomas Cleary in The Lost Art of War, by Sun Tzu II, San Francisco: Harper, 1996, p. 101.

CHAPTER

4

Contents

Recommended QMS Documentation

4.1 Overview of Documentation Requirements

4.1.1 Introduction

In Part 1, our goal was to create an organizationwide, business-oriented QMS, in which the financial and quality objectives were transparent. The QMS was to be based on the canonical set of ISO 9000 texts (i.e., ISO 9000:2000, ISO 9001:2000 (the Standard), and ISO 9004:2000).

It is now necessary to establish the key components of an effective QMS in terms of the Standard's documentation requirements, both from a mandatory basis and an implied overall effective hierarchy of documentation. In this regard, all documentation requirements (SHALLS) are to be addressed. Of prime importance are the mandatory documentation requirements, summarized in Section 4.1. These requirements are explicitly required by the Standard and form the umbrella under which all the other documents are contained.

To accomplish this, it is necessary first to categorize the several sets of documentation needed to produce a fully compliant and effective QMS. The four key sets are as follows:

- The Standard's mandatory documentation;
- The Standard's implied documentation;
- The registrar's required documentation;
- Required regulatory (compliance) documentation.

Whereas the Standard's mandatory documentation is defined by the Standard's SHALLS, there is considerable disagreement over what constitutes the various other required documents.

4.1.2 Recommended Documentation Taxonomy

The so-called ISO 9000 tiers (hierarchal levels of information) originated out of industrial-military requirements and have become a de facto standard because of their usefulness. The most common set of tier documents observed consists of the quality manual as tier I, SOPs as tier II, work instructions as tier III, and records as tier IV.

This set of tiers used by many writers in ISO 9001:1994 documentation systems has caused some confusion because forms were normally either missed or mixed in with records, even though form is a valid taxonomy term. By contrast, records are filled-in forms and can exist at any level of the documentation system. In fact forms, as structure, represent the lowest level of information flow and should be placed at the lowest level in the hierarchy. In addition, SOPs and work instructions are classified as different levels of documentation even though they are both procedures.

To complicate the matter further, SOPs are actually written in the form of both policy and process rather than procedure, as a procedure tells an individual how to do a specific task and is not meant to define a complex flow of information between functions. Unfortunately, the term *procedure* is still used in the Standard to include the description of a process, and so the confusion continues. A dictionary on this subject will not help because terms such as process and procedure are thrown together in a hodge-podge of equivalency. Something much more useful must be done to define an effective documentation taxonomy.

For the sake of clarity—which may be ephemeral—a process, as defined by the Standard's vocabulary, is meant to describe how a set of inputs is transformed into a set of outputs. The process moves through various phases until the activity results in a specified output.

On the other hand, to clearly differentiate between the defined process terminology, it is necessary to place a more constrictive use on the term *procedure*. For our purposes, a procedure describes the manner in which specific input activity, transformation activity, or output activity is accomplished. It is a subset of the process and is taken in steps that result in the completion of a specific activity (e.g., the Internal Audit Procedure describes the steps taken to carry out an internal audit, the In-House Calibration Work Instruction describes the steps needed to calibrate micrometers using a secondary Standard such as a gage block set, and a Power Supply Test Plan describes the steps needed to test out a power supply unit under robust conditions).

By contrast, for example, the Audit Process includes the previously described Audit Procedure plus all of the supplemental audit activities. Such activities include audit responsibility and authority, audit plans, audit training, audit corrective action protocols, audits of suppliers, audits by the registrar or regulatory agencies (e.g., FDA, notified bodies), and audit reviews by top management.

4.2 The Four-Tier Pyramid Concept

Our first task, then, is to remove the previously stated ambiguities and describe clearly what constitutes the general set of QMS documents so that our model is directly applicable to an effective documentation structure. As a memory aid, we will use the tier concept (i.e., tiers I–IV), but with a more specific set of definitions.

A useful icon in this regard is to place the four tiers in the form of a documentation pyramid (see Figure 4.1) [1].

Figure 4.1 The four-tier operational pyramid concept—ISO 9001:2000 guidelines.

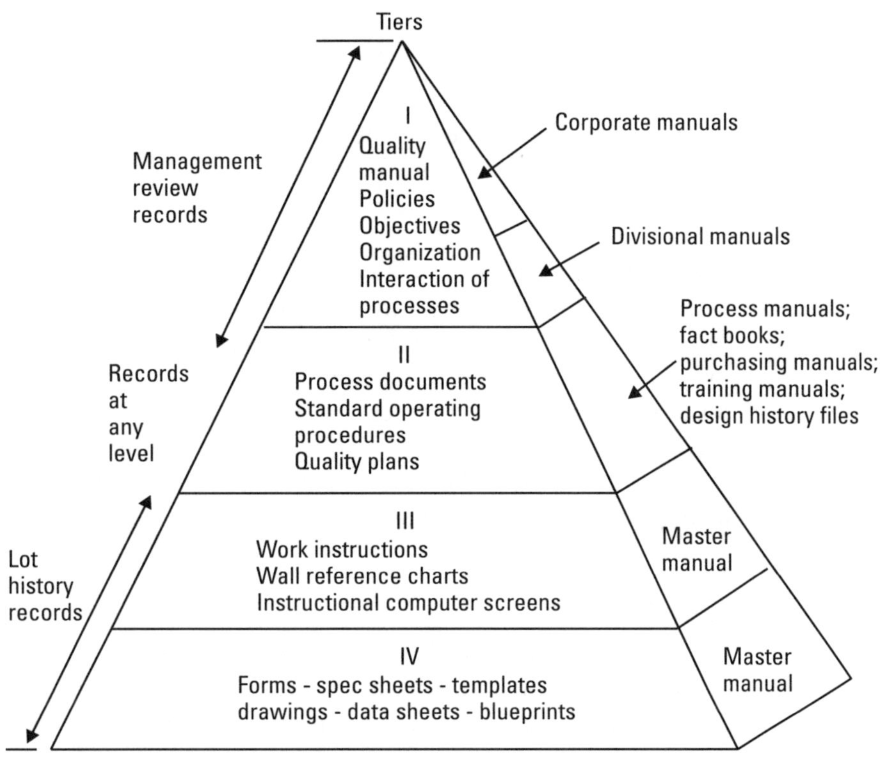

4.2.1 Matrix Format

The four tiers can also be described in the form of a table (e.g., see Table 4.1). The matrix form provides another class of information related to the specific content of a given tier [2].

4.2.2 Operational Tiers

We have specified in Table 4.1 that the documentation pyramid represents the operational flow of information (i.e., day-to-day processes carried out by use of dynamic and current documentation). That is why records are not included in tier IV. This is contrary to common usage, which we believe is incorrect from a taxonomy standpoint. We realize that this is a fine point, yet it causes considerable confusion among QMS designers.

Records should be listed as a distinct category of documentation within the QMS documentation umbrella. Although it is unnecessary to consider records in the form of a taxonomy, it is sometimes quite useful for organizational purposes. For example, the following records could be filed according to tiers:

- Management review minutes are records at the tier I level because they are part of the policy-making top-management control system.
- Corrective action reports are tier II records because they are directly associated with a SOP.
- A completed/filled-in tier III work instruction (e.g., verification test instruction), becomes a tier III record.

4.2.3 Guidelines

The proposed documentation taxonomy—policy, process, procedure, form—fits readily into this documentation pyramid. However, the tiers and various examples of documents are merely guidelines. It is the quality manual, quality objectives, identified processes and their controls, control plans where applicable, six specific procedures, supplemental documents if applicable, work instructions if applicable, and records that are clearly mandatory hierarchal documents in the Standard.

4.2.4 Four Tiers

The four-tier operational pyramid does emphasize the impact of the quality manual (manual) on the entire documentation structure, although—as noted—the pyramid is only meant to be a guideline because it does not

Table 4.1
The Four Suggested Operational Tiers of ISO 9001:2000 Documentation (Records Can Be Maintained at Any Tier)

Tier	ISO 9000 Category	Content Description	Deals with...
I	Quality manual Corporate Divisional Departmental	A time-independent document describing the organization's policies written in conformance with the Standard. Scope of QMS Details of exclusions Documentation of quality policy Documentation of quality objectives Description of organization Identification of processes Description of processes interactions Inclusion or reference of procedures	The organization's response to each SHALL The "rules of the house"—the methods used to ensure compliance Definition of responsibility
II	Process documents and high-level procedures SOPs Departmental operating procedures Business plans Quality plans	Time-dependent documents that describe either the overall processes of the organization or a combination of process and high-level procedures Enterprise processes Six mandatory procedures Documents needed to ensure the effective planning, operation, and control of the processes Employee handbook	Purpose—what, when, where, who, and why at a high level Flow of information from area to area, department to department, building to building
III	Lower-level procedural documents Wall reference charts Instructional computer screens Work instructions Directions	Time-dependent, detailed step-by-step work instructions on how to complete a task (e.g., at the operator or bench level)—sometimes integrated into tier II documents Purchasing work instructions Manufacturing work instructions Training syllabus	How one does the job—tells the reader in a step-by-step fashion Provision of the necessary data to perform the tasks
IV	Unfilled-in forms, graphics, or spec sheets Templates Blueprints Schematics Specifications Drawings	Generally time-independent documents that specify the data requirements called out in the various documents and/or specific data sources, or graphically indicate requirements or state specifications Many of the forms are used as records once they are filled in and filed, although specific records are required at all levels Complementary documents to support work instructions	The forms used to demonstrate that a procedure requiring either data taking or data input was done Drawings and/or specifications used in manufacturing or troubleshooting The templates required to measure and fabricate

replace the actual linkage that must be present from document to supplemental document.

4.2.5 Navigation Is Key

The four-tier structure readily provides levels for the type of documents that are usually encountered. However, some companies have as few as two defined levels and some as high as six defined levels. The number of levels is irrelevant. What is relevant is that they are presented in a way that aids the reader to easily navigate throughout the system.

4.2.6 Clearly Link Lower Tiers from the Manual

What is usually found is that the ISO management representative is cognizant of the total documentation structure but everyone else has great difficulty locating specific documents within the overall taxonomy. As a result, during an interview, when the auditee becomes confused over where to locate a document, I always suggest that they start with the manual and work down through the documented system. This usually helps, but only when the manual is clearly linked to the lower tier documents.

For online systems, this is readily accomplished with hyperlinks. However, it must be made relatively simple for the reader to quickly find the manual icon on the network, and with it the links.

4.2.7 Waterfall Effect

The use of a four-tier pyramidal structure for the QMS documentation is recommended to maximize communication to users. Once the four-tier hierarchy has been established, the total documentation system tends to behave with a waterfall effect (i.e., the number of process documents are less than the number of procedural documents, which in turn are less than the number of forms). We have illustrated this effect graphically in Figure 4.2.

In this figure, we have made the assumption that the quality manual is a stand-alone document (i.e., only contains quality policy statements and refers to lower tier documents). As demonstrated later, this is not the only possible configuration for the system, but it greatly helps to describe our concept.

The tendency for documentation growth must always be challenged. However, the use of the described techniques will tend to minimize this growth.

4.2.8 ISO 9000 Hierarchal Drivers

In Figure 4.3, we see that the four-tier concept is universal (i.e., the Standard defines the quality manual responses, the quality manual responses confine

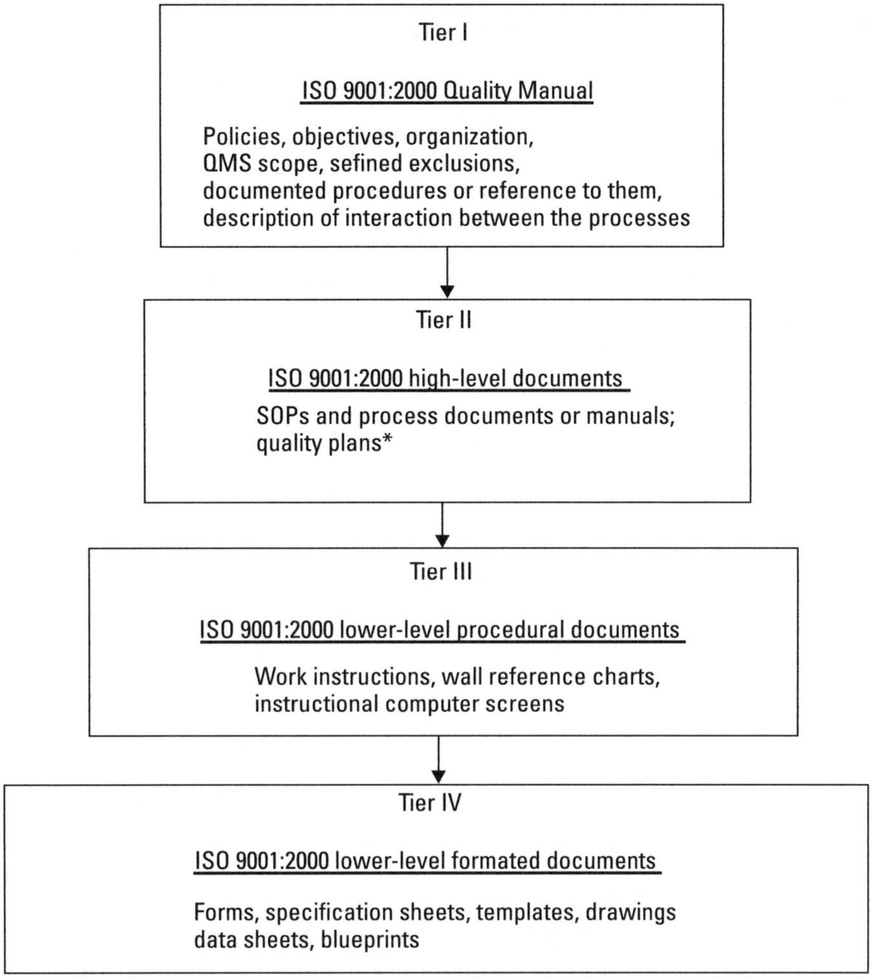

Figure 4.2 ISO 9001:2000 documentation waterfall effect [3].

*The terminology for tier II documents varies widely over different industries. We have found that SOPs, quality plans, process documents, and even low-level procedures fulfill the role of tier II documentation (i.e., that to which the tier I document sends the reader is intrinsically a tier II document).

the content of the second-tier documents, and the tier II documents drive the content of the procedures/work instructions). In this manner, the executive rules are transformed into management controls that are then transformed into operational directives.

Thus, the design of an effective QMS is holistic in that it is more than the sum of its parts. Unlike the engineering design of a personal computer's printed wiring assembly, there must be a powerful motivational element present within the QMS environment. For example, there is no need to motivate the electrons to flow efficiently within the printed circuit board's copper tracks, but there is an extremely important requirement to create a symbiotic

Figure 4.3 ISO 9001:2000 hierarchal drivers.

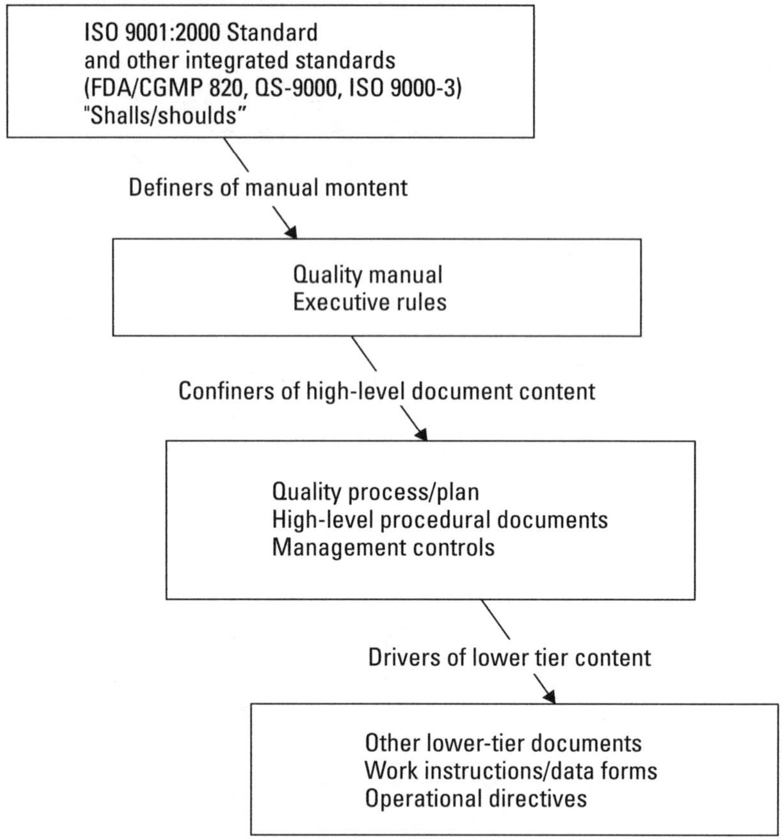

relationship between the inert document's pages and the dynamic application of those documents by human beings. Thus, there is always an affective requirement as well as an effective requirement in the design of the QMS.

The QMS acts as a living organism, and this is why it is so difficult to create the QMS in the first place and then to effectively maintain the system. However, it is the inherent ability of the four-tier structure to enhance informational flow that increases the probability of a successful QMS [4].

4.3 The ISO 9001:2000 QMS Is To Be Documented

The Standard demands a documented QMS. Within this mandated documentation are to be found the means to do the following [5]:

1. Identify QMS processes;
2. Determine process sequence and interaction;

3. Determine operational and monitoring criteria;
4. Determine operational and monitoring methods;
5. Monitor processes;
6. Measure processes;
7. Analyze processes;
8. Achieve planned results;
9. Achieve continual improvement of such processes.

So the key question here is how do you document these nine mandatory requirements? In other words, how do I document identified processes? How do I document the sequence and interaction of these processes? Our answer is to use a tier II document that can be either contained within the quality manual or referenced to another text from the quality manual. An agreement that this process document is a viable response to these nine requirements will validate the use of the proposed four-tier documentation structure.

With this assumption in mind, it is now necessary to establish that the following documents are the desired complete hierarchal set of documents, either defined or implied, in the Standard:

1. A documented quality policy (tier I);
2. Documented quality objectives (tier I);
3. A documented quality manual (tier I);
4. Six specifically defined documented procedures (tier III);
5. Documents that ensure the effective planning, operation, and control of processes (we logically conclude that this implies):
    - Process documents/SOPs/quality plans (tier II);
    - Procedures/work instructions (tier III);
    - Forms (tier IV);
6. Records (filled and filed forms that can occur at any tier level).

Of the six types of recommended global documents categorized here, records are the least understood with regard to their position in the documentation hierarchy. In fact, records (i.e., historical documents or documents used as objective evidence of activity) are distributed across the elements and can occur at any level of the hierarchy. For example, management review minutes are generated at the tier I level, quality plans and routers/travelers in the form of medical device history formats are required at the tier II level, and inspection and test reports are introduced at the tier III level.

To term records as being a specific tier is inappropriate. Records are a separate category and, in this regard, the Standard recognizes that that they are a special type of document and require document control [6].

4.3.1 Information Channel Management

In a more graphical sense, Figure 4.4 indicates the several possible channels of information that are covered under the suggested QMS four-tier concept [7]. Documents add up very quickly—a small company will reach hundreds of documents and a larger company will reach thousands of documents by the time they apply for certification. The number of documents will be proportional to the number of organizational functions and operating divisions. In addition, it is important to include controlled documents for field sales and field service personnel who are in residence outside of the main site's location but who must be kept up to date on revisions to the controlled documentation. Many companies now use online systems for this purpose.

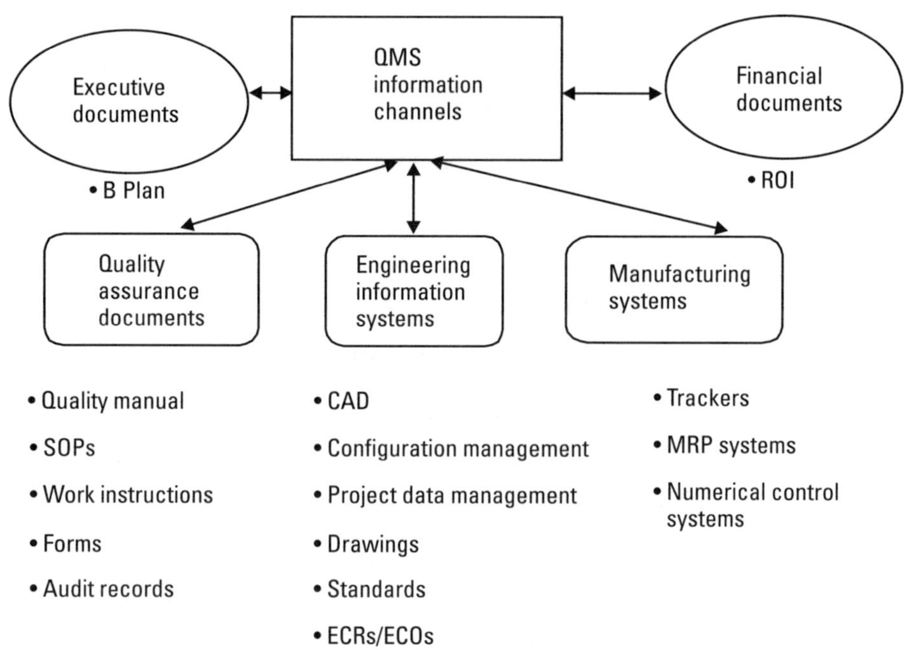

Figure 4.4 Potential QMS information channels.

4.3.2 Mandatory Tier II Linkage Requirements

Linkage is also defined in the Standard in that the quality manual is to either include the required procedures or reference them [8]. As a result, the tiers should be clearly linked so that it is possible to readily navigate throughout the documentation. One of the more effective ways to link documents is illustrated in Figures 4.5 and 4.6. In this case, the organization has used electronic media to control the quality manual. This online document exemplifies the nature in which the lower tier documents can be linked via hyperlinks and the various sections within the quality manual itself can be linked via bookmarks.

In practice, it is common to find the QMS defined primarily by the Quality Management Documents channel, with the Engineering Information Systems and Manufacturing Systems channels weakly described. The day-to-day

Figure 4.5 David Wayne Industries online quality manual cover page.

QMS Navigation Linkage

First Time Users — Begin with tier I, quality manual table of contents: TOC TOC

Expert Users — If you wish to directly view the other informational tiers of the QMS first, click the hyperlinked titles to move directly to the appropriate master lists to acquire the necessary documentation:

Tier I – quality manual table of contents TOC
Tier II – process manual ..\ISO DOC T2 \Master List.doc
Tier III – procedures manual ..\ISO DOC T3 \Master List.doc
Tier IV – forms master list
..\ISO DOC T4 \Engineering Forms \Master List.doc
..\ISO DOC T4 \General Forms \RECMSTER1.doc
-Records master list
..\Records\RECORDS MASTER LIST.doc
..\Records\DeptMinutes\MASTER LISTMinutes.doc
-Corrective action master lists ..\CAR \CAR Master List.doc
-Preventive action master lists ..\PAR \PAR Master List.doc
-Nonconforming material master lists ..\NCMRs \NCMR Master List.doc
-Supplier CAR master lists ..\SCARs \SCAR Master List.doc
-Reports master lists
..\Reports\Executive Master List.doc
..\Reports\Marketing & Sales Master List.doc
..\Reports\Engineering Master List.doc
..\Reports\Operations Master List.doc
..\Reports\Quality & Regulatory Assurance Mast er List.doc
..\Reports\Customer Service Master List.doc
..\Reports\MIS Master List.doc
..\Reports\Financial Master List.doc
-Audits master list ..\Audits\Audit Master List.doc
-Internal communications master list ..\Internal Communications \MASTER LIST.doc
-Quality objectives–metrics and charts ..\Quality Objectives \MASTER LIST.doc

Figure 4.6 David Wayne Industries online quality manual partial table of contents (TOC).

ISO 9001:2000 Quality management systems
Table of Contents

Manual sections	Section titles	Page or hyperlink
--	Cover page	--
i.	Table of Contents	--
1.0	DWI certification scope	S1
2.0	DWI vision	S2
3.0	DWI mission	S3
4.0	Quality management system	S4
4.1	General requirements	S5
4.2	Documentation requirements	S6
4.2.1	General	S7
4.2.2	Quality manual	S8
4.2.3	Control of documents	S9
4.2.4	Control of records	S10
5.0	DWI management responsibility	S11
5.1	Management commitment	S12
5.2	Customer focus	S13
5.3	DWI quality policy	S14
5.4	Planning	S15
5.4.1	Quality objectives	S16
5.4.2	Quality management system planning	S17
5.5	Responsibility, authority, and communication	S18
5.5.1	Responsibility and authority	S19
5.5.2	Management representative	S20
5.5.3	Internal communications	S21
.	.	.
.	.	.
.	.	.

interplay between engineering, operations, and quality assurance requires that all channels be equally efficient.

This was the reason, in the past editions of the Standard, that it was inefficient to certify a design and manufacturing facility to ISO 9002 and then seek a future audit to complete the certification to ISO 9001. The number of daily interfaces with engineering requires interface procedures. It is essentially the same effort to simply do the entire facility to ISO 9001 than to create all of those interface documents. The 2000 release resolves this issue because all certificates are to ISO 9001, and you are required to clearly justify why you have not complied with a specific clause required in Section 7, Product Realization.

The next step requires a careful examination of the Standard's mandatory QMS documentation requirements taken stepwise through the four tiers.

Endnotes

[1] We have just seen the hierarchy of policy, process, procedure, form used in the open literature: Bradel, Teri, "Quality Makes the Grade," *Quality Progress,* March 2002, p. 86.

[2] Although the terms *quality policy, process,* and *procedure* are defined in ISO 9000:2000, a pathological logic exists in the application of these concepts because a procedure is defined as a "specified way to carry out an activity or a process." It is a Catch 22 situation. Only six procedures are called for but we are to identify our processes, and it takes a procedure to document the process. As a result, our definitions are based on the work of Horn, Robert E., *Demystifying ISO 9000, Second Edition,* Information Mapping, Inc., Waltham, MA, 1994, pp. 5–6. In Mr. Horn's work, policy, process, procedure, and form are clearly defined.

[3] The terminology for tier II documents varies widely over different industries. We have found that SOPs, quality plans, process documents, and even low-level procedures fulfill the role of tier II documentation (i.e., that to which the tier I document sends the reader is intrinsically a tier II document).

[4] The importance of human interfacing with the QMS is extremely well documented. A source of original and lucid studies in this matter is available in the *Quality Management Journal,* a publication of the ASQ (e.g., Vol. 4, No. 2, 1997).

[5] ISO 9001:2000, Clause 4.1.

[6] ISO 9001:2000, Clause 4.2.3.

[7] The exponential increase of electronic media-based systems is concurrent with the explosion in information technology. It is now common to see electronic documentation control systems in use in parallel with MRP/ERP manufacturing control systems. Electronic calibration control systems are commonplace.

[8] ISO 9001:2000, Clause 4.2.2.

CHAPTER

5

Contents

Quality Manual Design

5.1 A Quality Manual Is a Mandatory Document

A quality manual is a document that specifies an organization's QMS [1]. The Standard further specifies that the quality manual is to include the following [2]:

1. The scope of the QMS and details and justification for exclusions;
2. The documented procedures or reference to them;
3. A description of the interaction between the processes.

And that's it—a broad umbrella requirement and only three specific conditions. Exclusions refers to a statement of nonapplicability for a particular clause. The most common exclusion is Clause 7.3 Design and Development (e.g., a machine shop that manufactures parts based on the customer's design would exclude Section 7.3).

The manual stands as a colossus above the ISO 9000 documentation hierarchy and sources the flow of the QMS. Accordingly, this chapter describes the design of a fully compliant ISO 9001:2000 Manual. However, although it is always best to keep the sections of documentation as small as possible, because of the approximately 40 specific areas of interrogation (i.e., distinct portions in the Standard with specific requirements), this is not possible. For example, Section 7 of the Standard has approximately 12 such distinct categories (i.e., 7.1, 7.2.1, 7.2.2, 7.2.3, 7.3, 7.4, 7.5.1, 7.5.2, 7.5.3, 7.5.4, 7.5.5, and 7.6).

As a result, we require an increased level of complexity beyond what we would normally consider. In fact, this book's longest text concerns the design of the manual. Such complexity requires increased employee training in documentation structure and usage. We will assume that the organization is committed to this relatively high level of employee training and has either already mastered or plans to master this art.

Unfortunately, the manual's dominant position in the QMS structure has not been supplemented by explanations that permit the rapid development of effective manuals. It is important that we first deal with the quality manual controversy as a way to understand how to resolve the issue.

5.1.1 The Manual Should Be User Friendly

As a result, the manual should tell the reader at least the following structural information [3]:

- The number of tiers chosen, their contents, and the method of labeling (see Figure 4.1);
- The method of document-to-document reference, or linkage (e.g., by either reference numbers in the text or a documentation tree);
- Whether the system is hard copy, on electronic media, or a mixture;
- The type of documents to be found (e.g., found either in manuals, in online documentation, by individual copies distributed among employees and/or locations, or on wall reference charts in which the work instruction is posted on or near the work station;
- How quality management documents (e.g., the manual) are differentiated from engineering documents (e.g., drawings and schematics);
- The identification of processes and the way that they interact, complemented by a description of their interaction.

Furthermore, we are required to specify a QMS, but the word specify is not defined. We are required to include the scope of the QMS, but the word scope is defined only in a guideline as "the range or extent of action, main purpose, intention"—that leaves much to the imagination [4]. Hence the need for guidance in terms of hierarchal structure and the need for concepts such as prescriptive quality policy statements. In this regard, we recommend that the quality manual should also include the following:

1. The documented quality policy and its mandatory requirements (refer to Par. 5.3 of the Standard);

2. The quality objectives and its mandatory requirements (refer to Par. 5.4.1 of the Standard);

3. The mandatory identification of the core competencies (processes) and how they are applied (refer to Par. 4.1a of the Standard);

4. The mandatory description of the interaction between the core competencies (processes) (refer to Par. 4.2.2 of the Standard).

5.1.2 A Quality Policy Statement Is a Mandatory Document

We recommend that the quality policy statement be placed within the quality manual. This approach considers the need for ease of distribution and overall visibility. However, the quality policy could just as readily be a separate document. In fact, many organizations place the quality policy statement within the quality manual and then extract it for purposes of display and ready availability.

Either way, the document is to be controlled, usually by signature and date of the top manager, but it could also be signed off by the entire executive team and/or the entire set of employees. The Quality Policy Statement must contain all of the Standard's requirements in Par. 5.3. See the boxed text below for an example of such a statement entitled, "Excellent's Quality Policy Statement."

5.1.3 Statements of Quality Objectives Are Mandatory Documents

We also recommend that the quality objectives, along with their metrics, be placed within the quality manual for the same ease of distribution and overall visibility. By metric, we mean the specific method of measurement (e.g., the ratio of awarded contracts to contracts awarded to competitors, first pass yields in test, or the percentage of shipments made to schedule divided by total shipments). We do not mean by this that the actual data should be placed in the manual, only the metric. For proprietary purposes, the reader should be directed to a separate document in which actual data exists, if and only if they are required to view such data. Targets and goals are included in the data.

Table 5.1 illustrates the kinds of metrics that can be generated to support various quality objectives. An example of a possible quality objective related to that table is discussed in Section 5.1.4.

5.1.4 Example

Here, the metric will be the percentage of returned product (in warranty). This is the calculation based on the total number of units shipped that remain

Excellent's Quality Policy Statement
The Business of Quality

The Excellent Corporation is committed, at all levels of the company, to total customer satisfaction. To meet this commitment, we provide products and services that fulfill customer expectations, and provide quality at levels greater than that available from any of our competitors.

Our quality management system is based on the International Standard ISO 9001:2000, and we are fully committed to continually improve the effectiveness of our system by means of constant top management review and oversight. This level of attention is complemented by formal management reviews, internal audits, extensive training, and an aggressive corrective and preventive action program that includes cross-functional teams for root-cause analysis and problem resolution.

To ensure the integrity of our system, quantitative quality objectives based on operational metrics are established, monitored, measured, and reviewed by managers who are held accountable for their results. In addition, all of our employees are thoroughly trained in quality management methods and are supplied with the resources required to ensure that such methods are effective.

At Excellent, business objectives and quality objectives are synonymous.

Signed: *John J. Dewolf* Dated: January 1, 2002
President & CEO
The Excellent Corporation

in a 12-month warranty. The associated quality objective for manufacturing in flow-down is "The manufacturing department is committed to minimizing the number of returned parallel processor boards by means of a vigorous joint program with purchasing to improve overall first pass yield." The metric, percentage of returned product (in warranty) out of the total population of warranty boards, will be used to define the overall quality of the manufacturing/purchasing cross-functional team.

In addition, manufacturing will also use first pass yield as the internal metric and purchasing will use the percentage of acceptance of supplier parts at incoming inspection to monitor supplier performance as its metric. First pass

Table 5.1
An Example of Metrics for Hickory Electromechanical Engineering

Metric	Objective	Metrics and Goals	Responsibility	Databases
1	Maximize contracts awarded (by product)	Contracts awarded to us divided by contacts awarded to competitors (by product); goal is to achieve 100% of awards	Program manager	Data table ..\Reports\COc
2	Minimize returned goods (RGAs)	Products returned over products shipped; goal is to be at 1% or less of shipments	Sales support	Data table ..\Reports\RmaMetRpt.xls
3	Minimize customer complaints	All complaints analyzed; goal is to maximize positives and minimize complaints and the response time for resolution	All who have customer contacts; business management tallies	Data table ..\Reports\Cus
4	Minimize customer correction actions (CARs)	Corrective action metric; goal is to reduce over time the number of CARs received and to close CARs by due date	ISO management representative	..\NCR's\CARJs
5	Minimize vendor nonconformance reports	Nonconformance metric; goal is to maintain resolution with vendor at 100% to avoid escalation to a CAR	Purchasing	..\NCR's\NCMs
6	Minimize audit correction actions (ACARs)	Audit corrective action metric; goal is to reduce over time the number of ACARs generated and to close ACAR's by due date	Quality assurance	..\Audits\ACAIs
7	Maximize control of monthly financials	President's report; goal is to achieve budget levels or greater	Finance	..\Reports\XBt.doc
8	Maximize actual sales	Meet or exceed projected sales; goal is to achieve budget levels or greater	Sales manager	..\Reports\BudvActvF.xls
9	Project budget	Actual project cost; goal is to achieve project budget levels or better	Engineering	Data ..\Reports\XB.doc
10	Minimize actual delivery times/dates	Contract delivery time/dates; goal is zero days or fewer.	Operations	Data ..\Reports\XB.doc
11	Minimize factory acceptance test (FAT) deficiencies	Maximize yields; goal is 0% of FAT deficiencies	Production supervisor	Data table ..\Reports\FAc

yield is based on the multiplicative yields through incoming inspection yield multiplied by subassembly functional test yield, multiplied by systems validation testing yield (customer supplied test fixture).

The goals for the minimization of returned product are as follows:

- Percentage of returned product (in warranty) is less than 1%/month;
- First pass yield is greater than 85%;
- Percentage of acceptance of supplier parts is greater than 97%.

The goals/targets are best kept in a separate document that is referenced in the quality manual. The reason for this is that the goals could change a number of times during the year as field data is obtained. Such data could cause revisions to the controlled quality manual. In addition, you would want to include the analytical details for the metric calculations in a separate document (e.g., a document entitled "Statistical Analysis Procedures").

5.1.5 Performance Rate

By the use of quality objectives and clearly defined metrics to measure each objective, a key tactic of our integrated QMS is to set the organization's performance improvement rate such that it achieves the organization's business/quality objectives and is attainable from a marketing sense.

We wish to drive the organization's performance at some continuous rate of improvement by means of a clearly defined QMS that is subject to our market imperatives.

5.2 The Quality Manual Controversy

5.2.1 An Issue of Content

In the attempt to create a fully responsive QMS, we have found that the content of the manual is the most controversial element in the creation of an effective ISO 9000 documentation and implementation system. It is the root cause of misunderstanding and misapplication of the Standard.

By contrast, procedural documentation and forms, whether effective or not, are more easily understood by those who either create or use them. The reason for this wide discrepancy is not mysterious. The vast majority of our language deals with how things are done, how they are resolved, or how things are put together and taken apart. We seldom speak in terms of policy or principle unless litigation is involved. Then we throw up our hands and turn the problem over to our lawyers.

In the case of manual design, it is necessary to carefully consider the manner in which the ISO 9000 elements are discussed with regard to policy; the degree of detail required to describe such policies; the location of policy within the manual versus a location in lower tier documents; the use of paraphrasing instead of clear, descriptive text; and the choice of presentation as either a stand-alone document or an integrated set of documents.

This issue of content existed for the 1987 and 1994 versions of the international Standard and has already shown to be the same for the 2000 version. In fact, there were ISO 9001:2000–paraphrased manual software advertisements on the market before the Standard's final draft was even released.

All of these topics are discussed in detail within this book and the reader is presented with a number of alternative approaches to the manual's content—all of which, we believe, are in full compliance with the Standard.

Our observations with regard to manual content are based on the review of over 100 manuals during 10 years of firsthand experience in the accredited certification of more than 100 companies ranging in size from eight to 2,000 employees—in industries as diverse as printed wiring assembly contract manufacturing, medical device houses, manufacturers of electronic systems, and the growth of laboratory-bred animals. We have also observed the same issue in the areas of QS-9000 for automotive and ISO 14000 for environmental, based on a much smaller firsthand sample.

This wide interpretive disparity of the ISO 9000 guidelines—especially the interpretation of the ISO 10013:1995 "Guidelines for Developing Quality Manuals"—has proven to be counterproductive for both suppliers and ISO 9000 practitioners. Unfortunately, the clients have been placed in the middle of the conflict and have become a captive audience who must agonize over what is best for their organizations based on controversial information. This situation is hardly conducive to the elusive customer satisfaction we wish to achieve.

5.2.2 Manual's Value

We believe that a clear controversy exists with regard to what constitutes a fully responsive manual and thus an effective QMS. The question about whether we should exert our efforts to resolve the controversy is not as clear. As a result, we first need to examine the value of the manual to the organization in the ISO 9001:2000 QMS and clarify its strategic role in both the certification process and the development of an effective QMS. That the manual is considered a significant document as part of quality management has been established in part through rigorous analysis [5].

This book asserts that it is through the flow of prescriptive, quality policy statements into the lower tier documentation—and the resultant operational feedback by users of that documentation to the organization's top management—that completeness, closure, and continuous QMS informational growth occurs.

5.2.3 Major Gate

As indicated in Figure 5.1, once the strategic plan is created and the stewardship established, the manual is the first critical documentation gate an organization must pass through to complete their QMS.

We note that the manual is derived from an analysis of the organization's total processes—from its strategic front-end core competencies to its after-sales service activities (e.g., from marketing and sales to engineering to manufacturing to the repair of returned product).

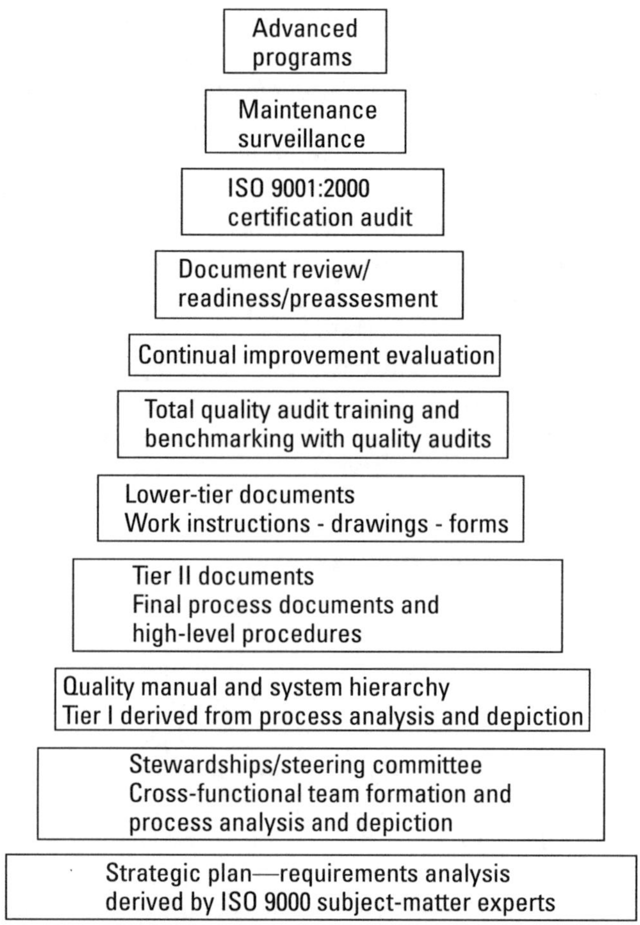

Figure 5.1 Typical ISO 9001:2000 certification gates.

Most importantly, to satisfy the ISO 9001:2000 requirements, the manual must include a description of the interaction between the several processes of the QMS. Thus, the manual becomes the unifier of strategic organizational thought. It is the glue that binds the various operational areas together. It is the first document that fully encompasses the manner in which the organization carries out its core competencies.

Once the manual is complete, the process/procedural documents can be finalized and a total quality auditing system can be put in place to monitor the effective implementation of the QMS. By total quality auditing, we mean audits at the system, process, and product level—internally via the organization's employees and also by customer and third-party auditors, and externally at the subcontractor's/supplier's facilities [6].

In addition, the manual is the first document seen by the registrar's auditors when the time comes to schedule the initial (certification) assessment. Unless the manual is acceptable, the registrar normally can go no further into the process. The manual review (a central part of the document review process) is often accomplished by the registrar as either an off-site activity or part of a documentation review in your facility prior to either the on-site initial assessment or the optional on-site preassessment.

The manual is also the primary document requested by your customer/client in their evaluation of your QMS. A fully responsive manual will often remove the necessity of an on-site audit, while a manual that simply repeats the Standard's text (a paraphrased manual) leaves your customer/client with no alternative but to visit you to better judge your conformance to the Standard if you are still considered a viable supplier. There are, of course, situations where the customer actually requests a trivialized manual. If that is what the customer wants, that is what the customer gets. However, that does not mean that the operational manual used by you should be the trivialized one.

The manual is the major driver for effectiveness in the QMS because it mandates the policies followed during management review, corrective and preventive action with customer complaints, analysis of data, and internal quality auditing.

5.2.4 Competitive Advantage

We have observed in practice that a fully compliant manual—that reflects both the personality and technical competence of the organization—significantly enhances the organization's competitive position. By comparison, we have observed that an inadequate manual has served as a competitive disadvantage and a source of delay in the certification process.

5.2.5 Rationale for an Ineffective Manual

In juxtaposition to the intrinsic value of a prescriptive manual, the following rationale is used by practitioners to answer why it doesn't matter if the manual is useless:

- Disinterest on the part of the customer;
- The only important documents are the operational ones;
- Disinterest on the part of the distributor;
- Protection of proprietary information;
- Disinterest on the part of top management;
- Wish to minimize the documentation effort;
- Insensitivity to those who want to understand the enterprise's strategic market position;
- Disinterest on the part of the employees;
- Misunderstanding of the manual's value as the primary navigator to find documentation;
- The position that the employees don't need to read the manual anyway;
- The customer simply wants a summary manual.

Of course, there are elements of truth in each of these positions. For example, selling product directly to either a high-volume end user or to a service provider can negate the need for a sophisticated quality policy manual. The warranty is of more importance. This situation, however, does not negate the need for the organization to have a dynamic internal QMS as a way to increase profitability through reduction of variance.

5.2.6 Conclusion

Based on this discussion, we see that although the manual is of an intrinsic value to the organization because it describes the strategic organizational viewpoint, a lack of definitive criteria establishes what constitutes a fully responsive manual. This interpretive issue has always existed with the previous ISO 9000 versions, and we already see a worsening situation for ISO 9001:2000, which requires even more interpretive aid than the previous versions [7]. The tendency is to continue to trivialize the manual through a slightly revised restatement of the Standard's text. We have concluded that the controversy needs much attention.

5.2.7 Observed Root Causes

We believe that the controversy is a result of rapid growth in the ISO 9000 industry. For example, approximately 10 years ago, in the very early days of United States entry into the world of ISO 9000, we found that the manual's structure was essentially a nonissue. We believe that this was due to the following:

- Relatively few suppliers that were large, multidivisional companies with established quality-assurance and control departments [8];
- The involvement of basically only high-tech organizations;
- The involvement of a limited number of registrars;
- Lead assessors shared similar quality-assurance backgrounds;
- A strong TQM influence;
- Basic quality programs formed from Mil-Q-9858A and FDA/GMP 820 standards.

We observed that the manual controversy grew slowly after 1994, and then in the period 1995 to the present accelerated into what we believe is a major issue. We have concluded that this change is due in part to the following:

- The explosion of candidates in small, medium, and large organizations in extremely diverse fields;
- Candidates in widely ranging levels of technology;
- A plethora of registrars and consulting groups and more on the way [9];
- A broad spectrum of lead assessors with varied backgrounds;
- Enhancements to TQM—for example, reengineering, quality function deployment (QFD), six sigma programs, enhanced Statistical Process Control (SPC), best practices management; business process engineering;
- The introduction of both integrated Standards—for example, QS-9000/ ISO/TS 16949 (automotive), ISO 14000 (environmental), TL-9000 (telecommunications), AS9000 (aerospace), FDA/CGMPs (USA) medical devices, Medical Device Directive (International); and a profusion of ISO 9000 guidelines.

We offer our approach to ISO 9000 QMS design in an attempt to harmonize such widely disparate perspectives. We feel that our design rules form a common ground for discussion and agreement.

5.3 Strategic Framework for the Manual

5.3.1 Unified Approach—Integration of Enterprise Strategy with Quality Management

We support the Standard's imperative that the manual should integrate enterprise strategy with quality management. This unification of enterprise strategy with quality management is accomplished by top management when it establishes quality policies and quality objectives that include the total organization's functions, (e.g., executive, marketing and sales, research and development, engineering, manufacturing, after-sales service, and the complementary functions of finance, quality assurance, and management information systems). This holistic approach (i.e., where the sum is greater than its parts) is inherent in the ISO 9001:2000 requirements.

This initiative can be further enhanced to include other related Standards that form the total organizational management system. For example, The American Society for Quality (ASQ) and the American National Standards Institute (ANSI) have declared that the quality management theme for the twenty-first century is to be Management Systems Integration—the integration of a QMS with both an environmental management system (EMS) and an occupational safety and health management system (OSHMS) [10].

Because the manual, if properly structured, can be readily modified to include the requirements of other associated standards (e.g., the manual becomes the enterprise/quality integrator of this total organizational strategy because it serves as the fabric upon which is imprinted the vision of the organization). The prescriptive quality policy statements that are the central tenets of the manual drive the operational processes, which in turn form the basis for TQM. This process forces every author to think deeply about the organization's mission and purpose.

5.3.2 Unified Business and Quality Policy

For example, a way in which business strategy can be integrated within the Standard's quality policy requirements is demonstrated in Table 5.2. In this table we have given examples of the manual's opening responses to the Standard's Clause 4.1, entitled "Quality Management System." The several suggested paragraphs are presented as quality policy statements that incorporate a fully integrated business/quality management system approach. The examples are not meant to be a complete response to Element 4.1 but they do form a context that considers the interaction between the Excellent Corporation and its registrar.

The blending of quantitative marketing objectives and financial metrics into the business/quality objectives creates a manual that unifies business and quality strategies into one. In this manner the manual becomes the organization's repository of operational knowledge that can form the basis of a learning organization.

Quality management systems based on the 1994 version of the International Standard tended to ignore such key enterprise protocols as the manner

Table 5.2
Unified Business and Quality Policy Format—Typical Quality Policy Statements Against the Requirements of the Standard That Also Considers the Registrar's Requirements

Manual's Paragraph Labels and Content	Typical Paragraph Content in the Form of Quality Policy Statements
Business and certification scope (requires registrar's review and acceptance) (usually placed prior to Section 4.0 in the manual)	Describes history, products, and locations of the business covered by the certification assessment. Example: "The Excellent Corp. maintains an online documented and continually improving QMS that facilitates the sales and marketing, design, manufacture, and servicing of SMT assemblies worldwide. The corporate office is located in Boston, MA, and is supported by sales and manufacturing facilities in the United States, Europe, and the Far East."
Vision statement (usually placed prior to Section 4.0 in the manual)	Defines very-long-range business objective. Example: "The Excellent Corp intends to maintain a dominant and globally recognized market position in the SMT industry."
Mission statement; can be corporate level and/or divisional/department level (usually placed prior to Section 4.0 in the manual)	Defines key objectives required during the next several years to achieve the vision. Example: "To achieve dominance, the Excellent Corp. will continually improve the effectiveness of its processes based on the eight management principles inherent in the ISO 9001:2000 requirements."
Quality-policy statement (refer to Par. 5.3 of the manual for detail)	A relatively short thematic statement that embodies the basic quality principles that every employee can remember. A more complete quality policy statement is also prepared, signed by the top manager, and posted throughout the organization. Example: "Quality within the Excellent Corp. means never being satisfied with anything less than a delighted customer."
Quality objectives/metrics (refer to Par. 5.4.1 of the manual for detail)	A list of key measurements that are used to define organizational success. Example: "Continual improvement is measured through the trend analysis of (a) customer satisfaction and dissatisfaction (customer returns and complaints, reorders, overall market share); (b) internal improvement metrics (yields, scrap); (c) corrective and preventive actions; and (d) return on net worth."
Process-based QMS	Discusses the method used to describe the organization's processes and their interrelationships. Example: "Business and quality are synonymous at the Excellent Corp. and are inexorably bound by means of critical process maps of the organization's core competencies (i.e., marketing and sales, engineering, manufacturing, service, and finance). An annual budget based on business/quality objectives is used as the means to supply the necessary resources to effectively control these processes."

Table 5.2 (continued)

Manual's Paragraph Labels and Content	Typical Paragraph Content in the Form of Quality Policy Statements
Customer needs and expectations (refer to Par. 8.4 of the manual for detail)	Discusses how the organization obtains knowledge of customer requirements, satisfaction, and dissatisfaction. Example: "The Excellent Corp. determines customer needs and expectations by means of its core competency in marketing and sales, which is defined as a critical process map, and interfaces rigorously with the engineering core competency, which is also define by a critical process map. Formal surveys of customer satisfaction and dissatisfaction levels are routinely collected quarterly."
Propagation of the quality policy (refer to Par. 5.1 of the manual for detail)	Discusses how the total quality policy is communicated to all employees and who is responsible for its implementation and maintenance. Example: "The President of the Excellent Corp. holds an integrated business/quality quarterly meeting with the entire staff to review the organization's quality progress and status against its business/quality objectives."

in which price lists are created and approved, and the manner in that all marketing channels are a source of customer complaint inputs [11].

The powerful roles played by marketing and finance are no longer conjectural in the Standard but are to be expressed through the requirement to identify QMS processes and to define interprocess sequence and interaction. For example, although cost of quality has always required an explicit contribution from finance, the finance departments were usually considered outside of the "quality" requirements and finance personnel were treated as second-class ISO citizens. However, a few moments of introspection always produced the manner in which finance played a key role in the ISO process (e.g., cost-of-nonconformance estimates, engineering cost workups, pay-back calculations for new capital equipment, and scrap analysis). This obviously exclusive rationale should no longer be tolerated based on the Standard's all-inclusive requirements [12]. Indeed, the finance department has been rediscovered in ISO 9001:2000!

Furthermore, the manual, when installed on an intranet, can also be used as the basis for the organization's information technology (IT) imperative that supports process development—the cornerstone of enterprise reinvention that can result in customer delight. As such, the manual can provide the channel that ties together the global operations of an organization.

The use of electronic media and intranets has produced an impressive array of documented systems that support outstanding databases used to analyze QMS effectiveness. There is now sophisticated software to cover security, enterprise resource planning (ERP), payment systems, fulfillment, customer relationship management (CRM), direct procurement, indirect procurement, supply chain management (SCM), Web-based computer-aided design, web-based product development, partner relationship management, and the blossoming of wireless network technology [13].

5.3.3 Proprietary Information

Top executives are often concerned about the visibility of business metrics such as profit and loss statements and balance sheets. Such proprietary information need not be detailed within the manual. For example, marketing strategies, cash flow, and profit and loss information are readily placed in a separate business plan that is then referenced in the manual. Our concern is with process so what is most important is that there exists an effective protocol that stipulates how marketing strategies are to be developed, how cash flow is to be measured and controlled, and how profit and loss information is to be used to improve the operation's corrective and preventive action programs.

Any situation in which proprietary information could become an issue should be discussed ahead of time with the registrar (i.e., the organization you have hired to certify you). Usually, third-party lead assessors are very cautious about reviewing either proprietary business information or personal employee data. It is not uncommon for third-party assessors to be refused entrance to a specific operational area but to treat the area as a black box through the examination of what product or service goes in, what transformation occurs to those inputs, and what product or service comes out of the black box. This is readily achieved through audits of the pertinent documentation, records established in the restricted area, employee qualifications, and interviews with the local area managers held outside of the restricted area.

In a proprietary situation, there is always a way to review documents, test out implementation, and determine system effectiveness. For example, restriction from an area can occur in audits of laboratory-bred animals. I have held a number of audits speaking through a glass window to operators who held up records for me to look at in those areas that required more than a "bunny suit" due to contamination issues.

Thus, in the thoroughness of my work, I often come across specific information that the client does not wish to disclose. This is never a problem because my primary purpose is to analyze processes, and so as long as the process is properly documented in some way, is implemented according to some plan, and there is some way to determine effectiveness, it is unnecessary to report the proprietary details—only the verification and validation of the process.

As a result, a manual that stresses this unified perspective indicates that the supplier has considered ISO 9000 in terms of their overall strategic enterprise directions and has carefully determined how the QMS serves to support these directives.

In fact, we have observed a number of organizations that have chosen this unified approach as the starting point for their manual. In our opinion, it has resulted in a very effective document in terms of its value to decision-making readers. Such suppliers have received very high marks on their manual from both customers and registrars. Years later, even after several recertifications, the basic unified structure has so reinforced the QMS that it readily withstands major organizational changes including executive reorganizations, acquisitions, and the fluidity of the global markets.

5.3.4 The Design of Quality Policy Statements

The creation of effective quality policy statements required a decision with regard to the degree of responsiveness required in response to a given ISO 9001:2000 requirement (SHALL). It is necessary to agree on what is appropriate for our purposes.

The highest level of response is the manual, often termed tier I or level I, in that our response to the specific SHALLS is to be given in terms of quality policy statements (i.e., written declarative statements that explain how the organization conforms to the Standard's requirements) [14].

Ideally, a quality policy statement is in the form of prescriptive language that directly addresses the descriptive language of the Standard. For example, if the requirement mandates that a company's top management establish quality objectives that are measurable, the prescriptive response would be of the following type:

Wolf TL, Inc., measures its total business/quality performance by means of seven key metrics:

1. New product introduction time;
2. Bid wins versus loss ratio;
3. Surveyed customer satisfaction versus dissatisfaction;
4. First pass yield;
5. On-time versus late deliveries;
6. Rate of return of product;
7. Economic value added.

Such prescriptive quality policy statements drive the structure of the entire documentation system, the implementation of that system, and the manner in which the effectiveness of that system is demonstrated. Effectiveness is often seen as a reduction of variance throughout the enterprise's operating systems, which is reflected in increased productivity and profitability [15].

5.3.5 Manual's Value Within the QMS

We maintain the following:

- A QMS that is based upon a manual that is fully responsive to the Standard results in a strategic declaration of the organization's quality and technical competence as stated in the manual in the form of prescriptive quality policy statements. To be fully responsive means that the QMS has integrated business strategy with quality management in full compliance with the Standard's SHALLS;
- In opposition to a fully responsive QMS, we maintain that a *paraphrased* set of quality policy statements results in a less than effective QMS—by paraphrased, we mean a playback of the Standard's descriptive requirements in the manual—as opposed to prescriptive statements that indicate the methods used to actually conform to the Standard.
- Paraphrased manuals lack so much useful information about the organization that they are often simply ignored as a key document to review during internal quality audits.

5.3.6 Prescriptive Versus Paraphrased Methods

The QMS process and our thesis is illustrated in Figure 5.2. The diagram illustrates how the requirements of any Standard are heavily influenced by the organization's interpretive scheme to produce either a fully responsive QMS based upon prescriptive quality policy statements (QPS) in the manual or one based upon ineffectual, paraphrased statements in the manual.

The use of prescriptive quality policy statements explicitly drives the lower level process, procedure, and formatted QMS documents and produces a balanced set of hierarchal documents. The QMS is then dynamically balanced in terms of its documentation system, the implementation of those documents, and the degree to which one can demonstrate the effectiveness of the system to achieve its quality objectives.

By contrast, the use of descriptive, paraphrased statements produces a system that is narrow at the top and heavy at the bottom from a hierarchal

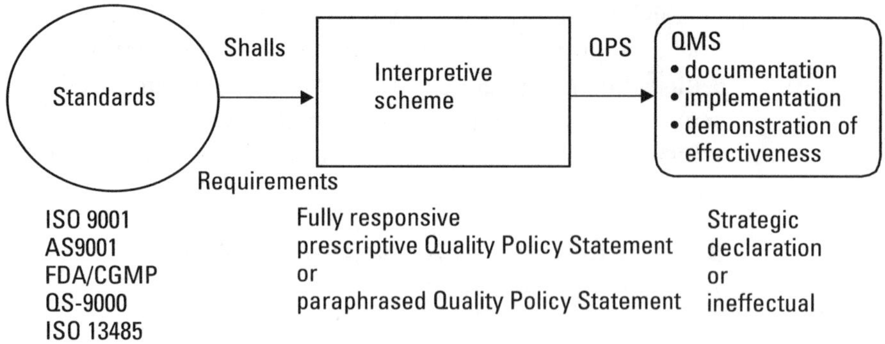

Thesis	Technique	QMS impact
ISO 9001:2000 shalls	Fully responsive >>	Strategic declaration
	Paraphrased >>	Ineffectual

Figure 5.2 The QMS manual thesis.

standpoint. There is no assurance in this case that the lower level documents accurately reflect an organization's strategic policies that are vaguely described in the paraphrased manual.

5.4 Cross-Functional Manual Action Teams

5.4.1 Section Experts

Most importantly, to produce an effective quality manual that has the capability to drive the entire QMS—that all employees can understand and relate to—requires that the stewards (champions) and authors have a detailed knowledge of the organization.

Each section of the manual is to be written by an author who has the most operational experience within that area (e.g., a technical or operational expert). It is one of the duties of the stewards to ensure that technical competence is optimized in this regard.

5.4.2 Ineffectiveness

Invariably, ineffective manuals result from authorship that does not take enough time to truly understand the organization's processes. This analysis of process is not only paramount in ISO 9000 but is just as critical in related activities such as reengineering and TQM.

We have often observed incomplete sections created by someone who is under a heavy time constraint and who is only remotely familiar with the section's content. It is the job of the steward to see that such situations are remedied through thoughtful assignments of personnel and a vigorous support system. It is unproductive to push the ISO schedule to the point where the organization's effectiveness suffers. There is no rush to certification—it is the usefulness of the created QMS that is important.

Just note how often certification deadlines are stretched out by those who demand them. The integrity of the business always comes first. The worst that can happen is that it will take a little longer and cost a few more dollars to complete the program.

5.5 SHALL Analysis

5.5.1 Definition of SHALL

The definition for SHALL is not located in the ISO 9000:2000 vocabulary, but is offered in the form of terminology guidelines [16].

The term SHALL is defined as follows:

> SHALL (SHALL NOT): Used to indicate a requirement strictly to be followed in order to conform to the Standard and from which no deviation is permitted. Do not use "may" or "may not" as alternatives.

Thus, both the SHALLS found in the Standard and the normative vocabulary definitions define the mandatory requirements for the Standard and establish the foundation for quality policy.

In a similar manner, the term *should* (*should not*) is also defined in the terminology guidelines and is used in guideline documents. It is "[u]sed to indicate that among several possibilities one is recommended as particularly suitable, without mentioning or excluding others, or that a certain course of action is preferred but not necessarily required, or that (in the negative form) a certain possibility or course of action is depreciated but not prohibited. (See ISO Directives Part 3:1997, Annex E.)" [17].

5.5.2 Appropriate Response to the SHALLS

Unless we answer the question, Within the QMS documentation system, where do we place the prescriptive responses to the descriptive SHALL requirements? we cannot define the nucleus of an effective documentation structure. As we noted previously, the Standard makes this decision neither obvious nor easy. A logical argument is required.

To reach a logical conclusion, we first need to examine what the Standard tells us about the quality manual. To begin with, there must be a quality manual. In addition, the quality manual must include the scope of the quality management system. Fortunately, the term *scope* is defined in the terminology guideline and agrees with common usage. For example, one dictionary's primary definition of scope is "1. extent or range of view, outlook, application, operation, effectiveness, etc.: an investigation of wide scope" [18]. The terminology guideline defines scope as "The range or extent of action, main purpose, intention." This definition implies that the quality manual is to contain a discussion of the entire structure of the QMS. We are then left with the question, "What is the intent of those who created the Standard with regard to what this discussion should look like and sound like?"

The determination of intent turns out to be the easier part of the argument. If we examine ISO 10013:1995(E), Guidelines for Developing Quality Manuals, Annex C, Example of a Section of a Quality Manual, we find that the committee's intent is to have each descriptive SHALL of the Standard addressed with prescriptive statements in the quality manual. This may not be your conclusion because the issue of the SHALL response remains in the realm of interpretation. However, in practice, many ISO 9000 practitioners support this conclusion inadvertently when they repeat/copy directly (paraphrase) the Standard with minor changes and declare that document to be the manual. Their act of paraphrasing includes every SHALL! This spirit with regard to prescriptive statements has carried over into ISO 9001:2000 with the addition of far more prescriptive statements than in the previous versions [19].

Although I abhor this blatant paraphrasing practice, it does indicate that my logic is acceptable as a reasonable resolution to the issue of SHALL response. In other words, if you don't think that we should respond to every SHALL, why do you copy every SHALL into the manual, even when they are not applicable? In short, if we can agree with this supposition, we have the nucleus for an effective QMS documentation structure. All we have to do is turn the duplicated SHALLS into prescriptive statements about what is really going on within the organization.

In this regard, we have found that the most intense areas of interpretive conflict center on the following:

- What needs to be expressed as policy (policy scope)?
- The level of detail expressed within the policy statements.

We will first discuss the scope of the policy statements.

5.5.3 Scope of Effort

Based on a relatively small sample of 12, it is already common to find ISO 9001:2000 manuals written in a sequenced form that corresponds directly to the Standard's sections and clauses (i.e., a manual with eight main sections and an appendix). All of the paragraph labels that deal with Sections 4 through 8 use the same nomenclature as the Standard so that there is a one-to-one correlation between the Standard's structure and the quality manual's structure:

- Cover pages/table of contents/document control;
- Section 1—History of the Enterprise;
- Section 2—Scope of the QMS Certification;
- Section 3—Quality Policy Statement;
- Section 4—Quality Management System;
- Section 5—Management Responsibility;
- Section 6—Resource Management;
- Section 7—Product Realization;
- Section 8—Measurement, Analysis, and Improvement;
- Appendixes.

Our rules require that Sections 4 through 8 should respond in detail to each SHALL within the corresponding element/clauses of each section. We now wish to determine just how many SHALLS there are so that we can estimate the scope of our documentation effort.

5.5.4 Effective Number of SHALLS

To estimate the number of SHALLS it is necessary to expand each SHALL statement in the Standard into all of its explicit and implicit requirements. The manner in which this is accomplished is discussed a little later in the text. As a matter of reference, we see in Table 5.3 that there are 135 explicitly stated SHALLS, and there are actually 364 expanded SHALLS (after the SHALL analysis) that require a response.

The expansiveness versus the Standard's explicitness is one of the more subtle and difficult parts of manual creation process. One often wonders, Where in the world did the assessor come up with that question? An experienced assessor

Table 5.3
Effective SHALLS per ISO 9001:2000 Section and Elements/Clauses

ISO 9001:2000 Sections	ISO 9001:2000 Elements	The Number of Explicit SHALLS	The Number of SHALLS After Expansion
4.0 QMS	4.1 General Requirements	5	20
	4.2.1 Documentation Requirements—General	1	8
	4.2.2 Quality Manual	1	2
	4.2.3 Control of Documents	3	14
	4.2.4 Control of Records	3	12
Subtotal 4.0		13	56
5.0 Management Responsibility	5.1 Management Commitment	1	5
	5.2 Customer Focus	1	1
	5.3 Quality Policy	1	7
	5.4 Planning—Quality Objectives—QMS Planning	3	5
	5.5 Responsibility, Authority and Communication	4	14
	5.6 Requirements—Management Review	5	17
Subtotal 5.0		15	49
6.0 Resource Management	6.1 Provision of Resources	1	8
	6.2 HR—Competence, Awareness, and Training	2	5
	6.3 Infrastructure	1	3
	6.4 Work Environment	1	2
Subtotal 6.0		5	18
7.0 Product Realization	7.1 Planning of Product Realization	4	15
	7.2 Customer Related Processes	8	21
	7.3 Design and Development	22	49
	7.4 Purchasing	9	18
	7.5 Production and Service Provision	14	41
	7.6 Control of Monitoring and Measuring Devices	9	17
Subtotal 7.0		66	161
8.0 Measurement, Analysis and Improvement	8.1 General	2	13
	8.2 Monitoring and Measurement	18	25
	8.3 Control of Nonconforming Product	6	9
	8.4 Analysis of Data	3	12
	8.5 Improvement	7	21
Subtotal 8.0		36	80
Totals		135	364

simply goes through all 364 expanded SHALLS specified in the Standard, one at a time, consciously or unconsciously. As you can see, there really is what seems to be an endless progression of interview questions. By the fourth or fifth surveillance, the depth of questions even surprises the assessor who introduces them.

Of course, the exact number of expanded SHALLS is not the point and the exact number is subject to all-night debates. It is the awareness of the completeness and underlying subtleties of the Standard that is significant. We believe, that the Standard, when fully responded to, establishes a powerful platform for the TQM scenarios discussed in this book.

As indicated in Table 5.3, this expansive nature of the Standard occurs in numerous elements. The use of this table becomes a useful guide when internal audit checklists are created.

For the sake of comparison, we note that the total number of expanded SHALLS for the 1994 version was calculated to be 320. It is interesting to note that a number of the additional 44 SHALLS of the Standard's calculated 364 occurs in design and development. For example, one of the new requirements is that the design output is to include information with regard to the purchase of production material, the actual production process, and the required after-sales activities, thus exemplifying the stress on total process not apparent in the 1994 version. The emphasis on customer satisfaction has also added additional SHALLS as part of the Customer Related Processes clauses. On the other side of the field, there has been a decrease of SHALLS in other areas (e.g., for the Control of Monitoring and Measuring Devices that tends to compensate for some of the newer SHALLS).

It is easy to get side tracked on this subject and miss the point that one is not to take the SHALLS at face value, but to look deeply into their implications and act accordingly. In this regard, we will now examine just how one can account for all of the requirements, explicit and implied.

5.5.5 Method to Count SHALLS

Because of its umbrella-like nature, we will analyze the first sentence of Part 4.1: QMS General Requirements of the Standard to demonstrate how we counted this expansive statement of SHALLS. Notice that there is only one explicit SHALL in the clause. However, also notice that the clause expands into five directives that require a response (not just one, that is). We are to do the following:

- Establish a QMS;
- Document a QMS;

- Implement a QMS;
- Maintain a QMS;
- Continually improve its effectiveness.

Because of its major importance in QMS structural definition, we will illustrate this technique further through the analysis of Clause 4.2.1: QMS Documentation Requirements—General.

4.2.1 Analysis In this case we have one explicit SHALL. However, when we expand the SHALL we get eight; that is, the QMS documentation is to include the following:

- Documented statements of a quality policy;
- Documented statements of quality objectives;
- A quality manual;
- Documented procedures;
- Documents needed to ensure the effective planning;
- Documents needed to ensure the effective operation;
- Documents needed to ensure the effective control;
- Records.

6.1 Analysis The analysis can get tricky. For example here is the SHALL analysis for Clause 6.1: Provision of Resources. (There is only one SHALL.) The organization is to do the following:

- Determine the resources needed to implement the QMS;
- Provide the resources needed to implement the QMS;
- Determine the resources needed to maintain the QMS;
- Provide the resources needed to maintain the QMS;
- Determine the resources needed to continually improve its effectiveness;
- Provide the resources needed to continually improve its effectiveness;
- Determine the resources needed to enhance customer satisfaction;
- Provide the resources needed to enhance customer satisfaction.

In this example, one SHALL yields eight expanded SHALLS.

This straightforward exercise, in practice, permits the reader to grasp for the first time the nuances of the Standard. What at first appears to be a document in which every sentence looks the same and is undifferentiated from any other sentence suddenly takes on the appearance of a good mystery novel as we start to look for the plot. We assume, of course, that you enjoy mystery novels. Better yet, it would really be helpful if you happen to be a Talmudic scholar.

5.6 Manual Section Length

There is some correlation between the number of expanded SHALLS and the number of pages in a given manual's section, but it is a weak one. The actual number of words required in a section has more to do with the scope of the SHALL, although Section 7.0: Product Realization is generally the longest because so many topics are covered, followed by section 4.0: Quality Management System because of the need not only to describe the system's structure but to also define the policies related to document and records control. Here we have assumed that appropriate quality policy statements have been created for each SHALL. A paraphrased manual—essentially just a play back of the Standard—will have sections approximately equal to the played-back Standard.

A good example of the inherent lack of precision in such an estimate is to consider that for a service organization whose specialty is repair and calibration of test equipment, Section 7.6: Control of Monitoring and Measuring Devices can represent over 40% of the manual.

Sections 4.0 and 5.0 require some graphics in the form of charts and figures that also increase the number of pages in those sections, although appendixes can be used to decrease the explicit size of the section. A graphic for Clauses 7.3: Design and Development and 7.5.2: Validation of Processes for Production and Service Provision is common if the system is online.

Font size differences make a study of such relationships almost meaningless. Instead say what is necessary, regardless of length, and use a font large enough to be read with ease.

5.7 Concomitance

Each SHALL denotes a specific requirement of the Standard. The requirements are often linked in such a way that they are associated (reciprocal, canonical) requirements. We have defined this associative characteristic of the clauses as concomitant relationships. As a result, if any one SHALL is not adequately addressed, there is an impact on other sections of the Standard because each SHALL is a part of the Standard's overall fabric.

As we have previously noted, each SHALL in the Standard is written to be descriptive with the intent that we are to reply in a prescriptive manner. By prescriptive, it is meant that the response includes operational details (e.g., what really happens, the actual method used, the specific approach taken in response to the requirement) but not the procedural details, unless you wish to produce an integrated manual that contains both policy and procedure. In either case, prescriptive statements are implied. In this way the manual can be used effectively by decision makers (e.g., customers who must decide on whether or not to audit their sub supplier based on a reading of the manual).

5.7.1 Requirement

Thus, the STANDARD's descriptive requirement, 4.1 General Requirements, might be responded to prescriptively as follows:

> Example response: "4.1 General Requirements: David Wayne Environmental Systems (DWES) has established a quality management system (QMS) that conforms in detail to the ISO 9001:2000 International Standard (Standard). The QMS encompasses the eight quality management principles described in the Standard. Document control and maintenance of the QMS is provided by the Director of Quality Assurance. The ISO 9000 Steering Committee, comprised of DWES top management and invited management representatives, oversees the effectiveness of the QMS via monthly reviews of the system's performance metrics. The objective of the Steering Committee is to ensure continual QMS improvement through the use of Quality Action Teams (QATs) assigned to resolve business nonconformances based on quantitatively arrived at solutions."

5.7.2 Training Example of Concomitance

Let us assume that prescriptive quality policy statements are in use and now look at an example of this concomitant—or interelement—relationship.

Our first example concerns training that occurs as follows:

- Explicitly in Clause 6.2: Human Resources;
- In Clause 7.5.2 (b): Validation of Processes for Production and Service Provision, where it requires that we determine that personnel are adequately qualified for their positions;

- In Clause 7.4.2 (b): Purchasing Information, where it requires specific training of the supplier's personnel when it makes sense to do so.

Training concomitance also occurs in the following ways:

- Implicitly in Clause 4.1(d): General Requirements, as it relates to resource availability;
- In Clause 5.1(e): Management Commitment, where again it refers to the resource availability;
- In Clause 5.6.3(c): Review Output, in regard to the determination of needed resources;
- Again in Clause 6.1: Provision of Resources, where we are required to decide on what level of resources are needed and then provide such resources.

Table 5.4, summarizes the various ways in which concomitance acts as a training binder throughout the requirements.

In each case, training is viewed from seven different, yet synergistic, perspectives:

1. Explicitly, in Clause 6.2, we discuss training as a top-down enterprise strategic issue.
2. Explicitly, in Clause 7.5.2(b), we consider the impact of training on the methods used to validate special processes (i.e., those that we cannot fully verify before the product is put into use by the customer).
3. Explicitly, in Clause 7.4.2(b), the possible training requirements for our supplier's personnel are considered.
4. Implicitly, in Clause 4.1(d), we again include training in an umbrella fashion to ensure that it is sufficient to successfully implement our processes.
5. Implicitly, in Clause 5.1(e), we admonish top management that its role is to ensure that adequate training is available.
6. Implicitly, in Clause 5.6.3(c), top management is again admonished to make sure that it checks on adequate training as part of management review.
7. Implicitly, in Clause 6.1, we are required to clearly determine and provide the training needed to employ an effective and customer-oriented QMS.

Table 5.4
Concomitant Training Relationships in Standard

Impact	Standard's Clauses	In That Clause We Are Required To...
Explicit	6.2 Human Resources	Indicate that all personnel performing work that affects product quality is competent on the basis of appropriate education, training, skills, and experience
		Determine just what that competence level is to be
		Provide for ways to support the competence needed
		Evaluate the effectiveness of the program
		Inform employees how they influence the attainment of quality objectives
		Keep records of such activities
	7.5.2(b) Validation of Processes for Production and Service Provision	Establish ways to validate special processes that includes the approval of equipment and the qualification of personnel
	7.4.2(b) Purchasing Information	Instruct our suppliers in the need for any specific personnel qualification requirements
Implicit	4.1(d) General Requirements	Indicate how the organization ensures the availability of resources and information that is required to support the operation and monitoring of its processes
	5.1(e) Management Commitment	Provide evidence of its commitment to the development and implementation of the QMS and to its role in continually improving QMS effectiveness by ensuring the availability of resources
	5.6.3(c) Review Output	Include any decisions and actions related to resource needs
	6.1 Provision of Resources	Determine and provide the resources needed to both implement and maintain the QMS and continually improve its effectiveness, and enhance customer satisfaction by meeting customer requirements

A more general perspective can be gained if we look at the total set of clauses at a glance as summarized in Table 5.5. As we have demonstrated in the training concomitance example, this table indicates other cases where an element of the Standard either clearly or implicitly references another element. Two other direct references, for example, are as follows:

1. When Clause 5.6: Management Review refers directly to Clause 4.2.4: Control of Records.

2. Clause 5.4.2: Quality Management System Planning specifies that the requirements of Clause 4.1 are to be met.

The table also alerts the reader to specifically referenced ISO 9000 guidelines that are provided as notes within the Standard.

Table 5.5
Standard's Concomitance at a Glance

Type of Concomitance →	Key Issue Involved	Explicit (Sent Directly to Clause)	Implicit (Implied Associated Clauses)	No. of Link(s)
ISO Clauses (1994) ↓				
4.0 Quality management system				
4.1(a) General requirements-QMS *(4.2.1)*	Process(es) scope,exclusion	1.2	7.1(b), 7.5.2, 7.6, 8.1, 8.2.2, 8.2.3	7
4.1(b) General requirements-QMS *(4.2.1)*	Process interact		4.2.2(c)	1
4.1(c) General requirements-QMS *(4.2.1)*	Effectiveness		5.6.1, 6.1(a), 8.4, 8.5.1	4
4.1(d) General requirements-QMS *(4.2.1)*	Resources		6.1	1
4.1(e) General requirements-QMS *(4.2.1)*	Monitor and measurement		8.2	1
4.1(f) General requirements-QMS *(4.2.1)*	Continuous improvement		8.5.1	9
4.2.1(a) General doc. requirements *(4.2.2)*	Q policy, Q objectives		5.3, 5.4.1, 5.1(c), 5.6.1, 6.2.2(d), 7.1(a)	6
4.2.1(b) General doc. requirements *(4.2.2)*	Q manual		4.2.2	1
4.2.1(c) General doc. requirements *(4.2.2)*	Documented procedures		4.2.3, 4.2.4, 8.5.2, 8.5.3, 8.3, 8.2.2	6
4.2.1(d) General doc. requirements *(4.2.2)*	Other docs.		7.5.1(b)	1
4.2.1(e) General doc. requirements *(4.2.2)*	Records	4.2.4	4.2.3	2
4.2.2(a) Quality manual *(4.2.1)*	Scope, exclusion	1.2	4.1(a)	2
4.2.2(b) Quality manual *(4.2.1)*	Documented procedures		4.2.1(c)	6
4.2.2(c) Quality manual *(4.2.1)*	Process interact	1.2	4.1(b)	2
4.2.3 Control of documents *(4.5)*	Records	4.2.4	4.2.1(e)	2
5.0 Management responsibility				
5.1 Management commitment *(4.1.1)*	Continual improvement		8.5.1	9
5.1(a) Management commitment *(4.1.1)*	Regulatory		7.2.1(c), 7.3.2(b)	2
5.1(b) Management commitment *(4.1.1)*	Quality policy		4.2.1(a)	6

Table 5.5 (continued)

Type of Concomitance →	**Key Issue Involved**	**Explicit (Sent Directly to Clause)**	**Implicit (Implied Associated Clauses)**	**No. of Link(s)**
5.1(c) Management commitment *(4.1.1)*	Quality objectives		4.2.1(a)	6
5.1(d) Management commitment *(4.1.1)*	Management reviews		5.6	1
5.1(e) Management commitment *(4.1.1)*	Resources		6.1, 6.2	2
5.2 Customer focus *(4.3.2)*	Customer satisfaction	7.2.1, 8.2.1	6.1(b)	4
5.4.1 Quality objectives *(4.1.1)*	Quality objectives	7.1(a)	4.2.1(a)	5
5.4.2 QMS planning *(4.2.3)*	Customer requirements	4.1	4.2.1(a)	6
5.6.1 Management review *(4.1.3)*	Quality policy, quality objectives	4.2.4	5.4.1(a)	6
5.6.2 Management review *(4.1.3)*	Audits/CAPA/ customer satisfaction		5.2, 7.2.3, 8.2.2, 8.5.2, 8.5.3	5
5.6.3 Management review *(4.1.3)*	Resources		5.1(e)	2
6.0 Resource management				
6.1(b) Provision of resources *(4.1.2.2)*	Customer satisfaction		5.2	4
6.2.1 Human resources *(4.1.2.2)*	Training/ resources		5.1(e), 5.6.3(c), 6.1, 7.4.2, 7.5.2(b)	5
6.2.2(d) Human resources *(4.18, 4.1.2.2)*	Quality objectives		4.2.1(a)	5
6.2.2(e) Human resources *(4.18, 4.1.2.2)*	Records	4.2.4		1
7.0 Product realization				
7.1 Planning of product realization *(4.2.3)*	General requirements	4.1		1
7.1(a) Planing of product realization *(4.2.3)*	Quality objectives		5.4.1	5
7.1(d) Planing of product realization *(4.2.3)*	Records	4.2.4		1
7.2.2 Review of requirements related to product *(4.3)*	Records	4.2.4		1
7.2.3 Customer communication *(4.3)*	Customer complaints		8.5.2(a)	1
7.3.2 Design and development*(4.4)*	Records/reg.	4.2.4	5.1(a)	2
7.3.4 Design and development*(4.4)*	Planning records	4.2.4, 7.3.1		2
7.3.5 Design and development*(4.4)*	Planning records	4.2.4, 7.3.1		2

Table 5.5 (continued)

Type of Concomitance →	Key Issue Involved	Explicit (Sent Directly to Clause)	Implicit (Implied Associated Clauses)	No. of Link(s)
7.3.6 Design and development *(4.4)*	Planning records	4.2.4, 7.3.1		2
7.3.7 Design and development *(4.4)*	Records	4.2.4		1
7.4.1 Purchasing *(4.6)*	Records	4.2.4		1
7.4.2 Purchasing *(4.6)*	Training	6.2.1		1
7.5.1(f) Control of production and servicing provision *(4.9, 4.15.6, 4.19)*	Work instructions		4.2.1(d)	1
7.5.2(d)Val. of processes production and servicing provision *(4.9)*	Records	4.2.4		1
7.5.3 Identification and traceability *(4.8, 4.10.5, 4.12)*	Records	4.2.4		1
7.5.4 Customer property *(4.7)*	Records	4.2.4		1
7.6 Control of monitoring and measuring devices *(4.11)*	Records/ customer requirements	4.2.4 7.2.1	Guidelines note 1—ISO 10012-1, -2	2
7.6(a) Control of monitoring and measuring devices *(4.11)*	Records	4.2.4		1
8.0 Measurement, analysis and improvement				
8.1(c) M,A,&I—General *(4.10.1, 4.20.1, 4.20.2)*	Continual improvement		8.5.1	9
8.2.1 Customer satisfaction *(new)*	Customer satisfaction		5.2	4
8.2.2 Internal audit *(4.17)*	Records/CAPA	4.2.4, 8.5.2	Guidelines note 2—ISO 10011-1, -2, -3	2
8.2.2(a) Internal audit *(4.17)*	Production planning	7.1		1
8.2.3 Monitoring and measurement of processes *(4.17,4.20)*	CAPA		8.5.2, 8.5.3	2
8.2.4 Monitoring and measurement of product *(4.10, 4.20)*	Production planning Records	4.2.4, 7.1		2
8.3 Control of nonconforming product *(4.13)*	Records	4.2.4		1
8.4(a) Analysis of data *(4.20)*	Customer satisfaction	8.2.1	5.2	4
8.4(b) Analysis of data *(4.20)*	Product requests	7.2.1		1
8.4(c) Analysis of Data *(4.20)*	Preventive action		8.5.3	1

Table 5.5 (continued)

Type of Concomitance →	Key Issue Involved	Explicit (Sent Directly to Clause)	Implicit (Implied Associated Clauses)	No. of Link(s)
8.5.1 Continual improvement (4.1.3)	Continual improvement		4.1(f), 5.1, 5.3, 5.6, 5.4.1, 8.1(c), 8.2.2, 8.4, 8.5.2, 8.5.3	10
8.5.2 Corrective action *(4.14.1, 4.14.2)*	Records	4.2.4		1
8.5.2(a) Corrective action *(4.14.1, 4.14.2)*	Customer complaints		7.2.3	1
8.5.3 Preventive action *(4.14.1, 4.14.3)*	Records	4.2.4		1

5.7.3 Application

The concomitance table is used when we create the manual or prepare an audit. In these two cases, we wish to know how interactive each clause is as we write text or prepare an audit and make references to other documentation. The number of links becomes a clue as to just how to cross reference our work. As we might anticipate, continual improvement has the most links, followed by several other topics that include processes and objectives.

For example, in carrying out an audit of management responsibility, it is necessary to not only audit all the SHALLS of Section 5.0, but also the requirements in Par. 6.1 and Part 8.5.1.

As demonstrated in the table, implicit requirements are a common thread throughout the Standard. As a result, it is relatively ease for interpretations to vary considerably among practitioners. In fact, each issue of the monthly publication *Quality Systems Update* (qsunews@aol.com) devotes a several-page discussion of interpretations by highly qualified practitioners. The International Automotive Sector Group periodically releases interpretations for the QS-9000:1998 Automotive Standard interpretations that also appear in Quality Systems Update. You will undoubtedly find more concomitance as you begin to work the process of conformance. Suffice it to say that the Standard's web of requirements is most complex and one should not take such implications lightly if the desire is to have a completely responsive QMS.

As a result, to ensure total system concomitance, we maintain that it is necessary to respond to every SHALL in a given Standard's element and to present this information in the manual.

5.8 Nonapplicability of Specific SHALLS

Aside from the issue of exclusion, where we might need to exclude key functions such as either design and development or after-sales servicing, in those cases where a specific SHALL is not applicable (i.e., not appropriate to the organization's structure), a discussion is necessary that clarifies the reason for its dismissal. This is contrary to the general feeling that one should just ignore the requirement. One can do this, but the loss is in the ability to clearly communicate enterprise policy. Consider a more positive approach to such issues.

Positive Approach to Nonapplicable SHALLS

Clause 7.5.3 Identification and Traceability In this case, our company is not required to maintain traceability by either our customers or by some industrial Standard under which we operate. A positive approach can be used to clarify the nonapplicability of the SHALL and to state what actually occurs operationally.

> Clause 7.5.3 example response: Although traceability is not a contractual requirement of FLS Enterprises, Inc., the company considers traceability an integral part of its quality mandate. As a result, all final products are serialized and all key components integrated into its systems are serialized. A record of all serial numbers is maintained by quality assurance as part of the system history record during final test and inspection.

As another example, consider a positive response to 7.5.5: Preservation of Product in that we are to preserve the conformity of our product during its delivery to the customer's dock. This clause implies that we are to extend product quality protection beyond the plant. A positive approach would be to state that, "The Excellent Corporation considers product quality paramount in our customer relationships and the level of protection of all product shipped is determined by quality-assurance laboratory testing to the appropriate international standard, as necessary." This type of statement puts the customer at ease and implies that you are someone they can negotiate with productively.

Contrast this approach with the alternative (negative) dogmatic statement "F.O.B. [free on board] is always established at the Excellent Corporation's dock." Now the customer will wonder what difficult type of negotiations they are apt to run into with this supplier who sees the world without shades of gray.

Clause 7.4.3 Verification of Purchased Product The same approach can be used in the case of Clause 7.4.3: Verification of Purchased Product in that our company does not perform source inspection and would rather not have our customer audit our suppliers. The positive response might be, "The Excellent

Corporation does not consider it necessary to perform source inspection on our vendor's products because they are qualified initially by a potential supplier survey and are required to provide certificates of compliance with each shipment."

The negative response might be, "The Excellent Corporation does not require source inspection." The customer will wonder just how you control your subcontractors—or do you?

5.9 Appropriate Detail Level

There is also considerable debate with regard to the detail required in a quality policy statement. We have found that the conflict lies in the belief that a lack of detail is an appropriate way to create quality policy statements (e.g., just paraphrase the Standard and add the company's name).

We wish to demonstrate that although the level of detail varies widely with quality policy statements, what is of prime importance is that enough detail be available so that the reader can use your described rules and methods to intelligently make business decisions that impact their organization.

5.9.1 An ISO 9000–Certified Vendor

Every day buyers and quality personnel make joint decisions on whether to add a new key vendor to their approved supplier list. In those cases where the new supplier is ISO 9000–certified, one of the decisions that can be made is to avoid the expense of a vendor audit and rely on the depth and scope of the supplier's quality manual for the decision on whether to add the vendor to the approved supplier list (ASL).

We have been there when the quality manual arrives and is so nebulous that it cannot be used as part of the decision process. The first response is to laugh about it and wonder how in the world they were ever certified. The final response is anger because now thousands of dollars and several precious days must be spent to run the vendor audit. The final blow comes when the audit is run. The vendor turns out to have an effective QMS in place—another case were the manual does not reflect the competency of the supplier—a very common event.

We wish to demonstrate that a carefully determined level of detail in a quality policy statement can be an appropriate way to decide on the effectiveness of the statement.

5.9.2 Example #1—On Work Environment

First Statement The following is a broadly stated quality policy statement in partial response to 6.4: Work Environment: "All Excellent Corporation employees shall comply with work environment procedures."

Second Statement This is also a quality policy statement on the same subject, but it is definitive: "All Excellent Corporation employees wear safety glasses and ear plugs before entering designated manufacturing areas and are supplied with anti-ESD-compliant work stations, as required (Re: EC Work Instruction 6.4.01)."

Example #1 Analysis The first statement is a philosophical directive equivalent to reading back (or paraphrasing) the Standard. It uses the future tense, so it is not clear that the rule is actually in place yet. It is a common paraphrasing technique that deflates the effectiveness of the Standard. The second statement is in the present tense and has sufficient detail to be clear to anyone reading it that this is what the organization requires—and it can be readily audited. This statement is presented in a way that is similar to the phraseology used in ISO 10013:1995 Quality Manual Guideline, page 11.

5.9.3 Example #2—On Control of Monitoring and Measuring Devices (Clause 7.6)

First Statement A paraphrased quality policy statement: "The Excellent Corporation shall determine the monitoring and measurement to be undertaken and the equipment needed to provide product conformance to requirements."

Second Statement A definitive quality policy statement: "The engineering and quality-assurance departments of Excellent Corporation jointly establish the degree of acceptance testing required on all products during engineering transfer of product to production via ECO control; and documents such tests in production on specification work order. All inspection, measuring, and test equipment purchased for this purpose is reviewed and approved jointly by the engineering and quality-assurance departments for the accuracy and precision required. A 4:1 rule or better is used, as applicable, to determine instrument accuracy."

Example #2 Analysis The first statement yields neither responsibility nor method. The second statement is very clear and provides the reader with the type of knowledge necessary to either choose the vendor or prepare a productive audit of the vendor. One gets the feeling of competence and completeness.

5.9.4 Example #3—On Internal Audits (Clause 8.2.2)

First Statement A broad policy statement improved somewhat over a paraphrased one: "At Excellent Corporation, internal audits shall be scheduled on the basis of how important the activity is to the company and how well it has performed in previous audits."

Second Statement A definitive policy statement prescribed in ISO 10013: 1995(E), page 11, Par. 4.17.4.2: "The scope of the audits is determined with regard to the importance of the activities in question and the knowledge of any existing or likely problems. The audit frequency is, at least, for quality system audits—once per year; for product quality audits—twice per year; for process quality audits—once per year. Audit plans are made up and documented once per year. Checklists are prepared as an aid."

Third Statement This one is a definitive policy statement that specifies responsibility and avoids specific numbers, which could vary widely during the year: "At Excellent Corporation, to ensure QMS effectiveness, all of the elements of the Standard are audited against every department in both the Evergreen Street and Fir Street facilities, as applicable, with all areas covered during each calendar year by means of monthly audits. The director of manufacturing establishes the audit plan, selects and trains the auditors, and issues the internal audit schedules—which include vendor audits—as required. The frequency with which a given department is audited is based on the results of both previous audits and events that have occurred during the interim. The total audit program includes systems, process, and product audits. All audits require checklists prepared prior to the audit by the auditors and approved by the lead auditor."

Example #3 Analysis The first statement is again philosophical and uses the future tense. The second and third statements are detailed, yet concise, and loaded with information about the company. The reader has little trouble visualizing the depth and scope of the audit program in either statement #2 or #3.

5.9.5 Suggested Rule

A slightly modified journalistic formula of who, what, where, when, how, and why is a suggested rule to keep in mind as you write quality policy statements. For example, in Example #3, third statement, we can apply "the five Ws and an H" as follows:

- Who = responsibility and authority = director of engineering and lead auditor;
- What = the activity = an internal audit of all departments against all appropriate clauses;

- Where = location of activity = in the two enterprise facilities;
- When = frequency of activity = monthly audits during a calendar year;
- Why = objective = to ensure QMS effectiveness;
- How = method used = systems, process, and product audits.

The more we define the five Ws and an H, the clearer the quality policy statement becomes. The clearer the quality policy statement, the higher the rate of information within the enterprise.

5.10 Level of Detail in Practice

In our review of over 100 manuals, we have found a tendency to avoid detail in the quality policy statement based on the belief that the manual should be of a certain size, with the result essentially being to throw out the baby with the bath water [20].

The inappropriateness of this belief in a specific length (size) is self evident when you consider that ISO 9000 manuals are written for organizations with from two to 10,000 employees. In a very large company, the organization chart appendix alone can be half as long as the manual for a company with eight employees.

Often, the policies end up in the procedural documents, so there is an awareness that the information is required, but there is also an issue of where it belongs. Sometimes, the quality policy statements appear nowhere, much to the chagrin of the ISO 9000 management representative who searches with great gusto and futility to prove the existence of a response that was believed to be too complex to place in the manual.

Sometimes they even end up in a work instruction or on a form. In fact, the concept of clearly written, informative, prescriptive, quality policy statement remains a fuzzy issue to this date. Practitioners still promote the idea that the manual can be written in a few days by merely paraphrasing the Standard [21].

5.10.1 Summary of Quality Policy Statement Attributes

In summary, we maintain that a quality policy statement should be as follows:

- A prescriptive response to every SHALL in the Standard;
- Present tense as opposed to future tense;
- Clearly expressed in simple declarative prose;

- Not paraphrased;
- Of whatever length and detail is necessary to define the organization's rules and methods.

We conclude that the level of detail in a quality policy statement should be whatever the SHALL demands in context with the methods used by the organization. Short policy statements are not necessarily effective and are often inappropriate.

5.10.2 Electronic Media Solutions

Fortunately, the advent of relatively low-cost electronically linked software is now available for document control for even small companies with small budgets. As a result, we would like to show some typical ways in which QMS designers have moved the suggested four-tier system onto electronic media via both internet and intranet information transfer systems.

Figure 4.5 illustrated how the cover sheet of the manual can be used as a central dispersion sheet for the entire QMS documentation. By opening up the QMS documentation via an icon that goes directly to the manual, each user can decide just how deep into the system they wish to go based on their expertise in navigating the various linkages. Those who are unfamiliar with the linkage would begin with the policy sections and those who know where they want to go will directly use the appropriate lower tier links.

For those who wish to search through the manual, there are links available on the table of contents pages so that the reader can link directly to any specific policy or process required. Once there, they are provided links to the next lowest tier. The internal book links were illustrated in Figure 4.6.

For example, we note that the text in the manual tends to be time-independent. These are statements of policy—rules of the house, methods, tools of use—that generally do not imply movement or process.

By contrast, the text in the lower level documents are time-dependent statements. The text in such documents generally imply or reference process or procedure as movement (e.g., from corporate to division, from division to division, from department to department, from operator to operator, or from operator step to operator step).

In the far right column of the table of contents, we also see comments that tell us how to deal with document content. As a result, if policy is presented in the manual, it need not be restated in the procedures. This is a common problem and policy shows up all over the several layers of documentation. Redundancy is to be avoided to minimize the number of documents that you will need to change when revisions occur, and they do occur frequently.

5.11 Pyramid for a Manual

In a similar fashion we can describe the hierarchal content of the manual as illustrated in Figure 5.3.

We indicate in Figure 5.3 that the manual contains the entire set of organizational quality policies (defined as phase 1). We have chosen to indicate five directly sequenced sections to cover the five operating sections of the Standard (defined as phase 2). This set of definitions is valid for any form of manual sequences or configurations.

Phases 3 and 4 are somewhat more difficult to define because they are parallel processes in that each SHALL of the Standard (phase 4) is responded to with a quality policy statement (phase 3). It is this four-phase process that transforms a descriptive ISO 9001:2000 requirement into a set of prescriptive quality policy statements.

We can clarify the language used in this graphic by a review of previous statements and definitions.

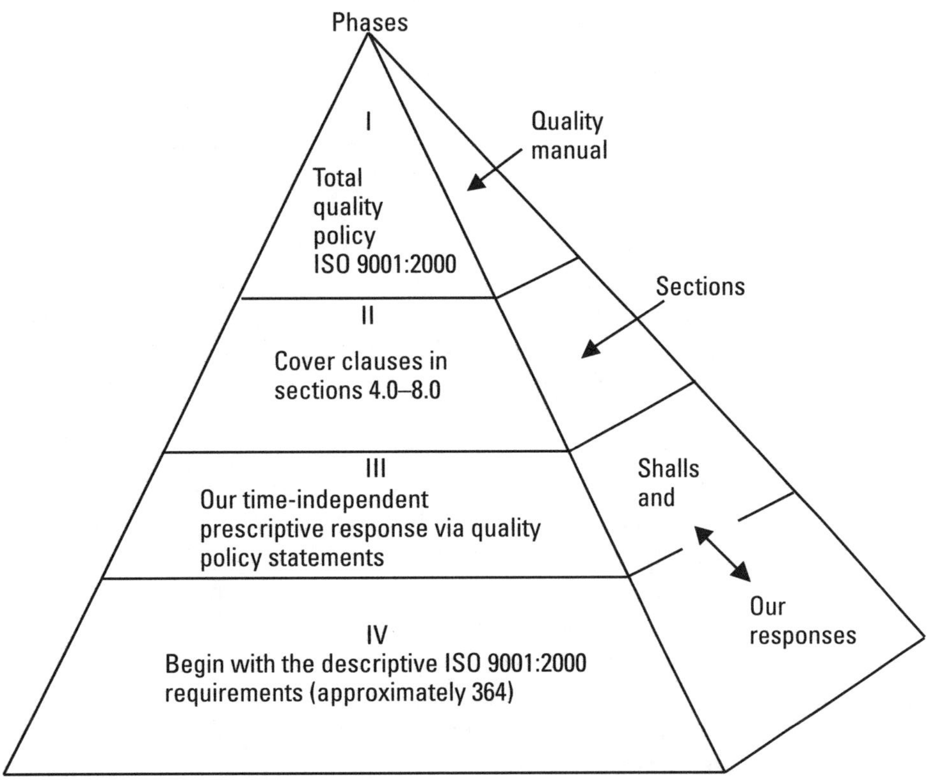

Figure 5.3 The four phases of the ISO 9001:2000 Quality Manual (direct sequence with Standard's sections).

5.11.1 Quality Policy

Policies are by their nature time-independent (i.e., they do not describe movement, but rather define position), whereas processes and procedures are time dependent (i.e., they describe flow, continuity, and movement). A policy is basically a rule of the house set up by top management. They are prescriptive (have specific direction and/or instruction) and indicate method of approach.

5.11.2 Total Quality Policy

The total quality policy consists of our response to all of the Standard's quality policy requirements as described in the quality manual.

5.11.3 Elemental Policies and Specific ISO 9001:2000 Requirements

There are five major sections in the Standard, which contain approximately 364 descriptive requirements in the form of either explicitly stated or implicitly directed SHALLS. In the case chosen, we have assumed that the requirements will be in the manual in five sections on a one-to-one basis with the Standard. More sections can be added as necessary (e.g., to meet regulatory, security, or safety requirements, although such requirements can be readily enclosed within one of the five sections).

All 364 requirements need to be addressed with prescriptive quality policy statements written into the text against their pertinent elements. Regulatory, security, and safety requirements would add more SHALLS to this number but are to be treated in the same prescriptive manner.

5.11.4 Quality Policy Statement Examples

We have maintained that each SHALL must be addressed if the manual is to clearly define the overall structure of the documented system and its effective implementation. We have previously discussed the requirements for documentation that are to be effectively implemented. This set of documents is somewhat complex and distributed throughout the Standard in Sections 4 through 8.

Thus, a quality policy statement is required by the supplier in response to each SHALL. The quality policy statement is intended to be prescriptive and to delineate authority/responsibility.

Some examples of quality policy statement are illustrated in Table 5.6. Appendix B addresses several more areas of the Standard not already addressed in the main text. Notice how specific each statement is and how authority and responsibility is clearly stated. In just these few sentences, it is possible to begin to visualize the Excellent Corporation's management structure and commitment to quality.

Table 5.6
Examples of Quality Policy Statements

Standard Clause	Standard's SHALL(s)	Quality Policy Statement(s)
5.2: Customer Focus	Top management is to ensure that we have determined and met customer requirements to enhance customer satisfaction.	The executive committee of the Excellent Corp. has assigned the vice president of marketing and sales to ensure that customer requirements are fulfilled—as described in the Marketing and Sales Process document—and that customer satisfaction is measured by means of several sales and marketing objectives, metrics, and goals summarized in the document entitled "Annual Business Objectives and Status." More discussion on this subject can be found in Sections 7.2.1 and 8.2.1 of this manual.
5.5.1: Responsibility and Authority (R&A)	Top management is to ensure that responsibility and authority for the various QMS activities are defined and communicated throughout the enterprise.	The executive committee of the Excellent Corp. has defined responsibility and authority by means of: An organizational chart (Re: Appendix A); Job descriptions (Re: Job Description Manual); Assignments summarized in tier I and II documents; Master R&A lists maintained by department managers.
7.4.1: Purchasing Process [last sentence]	The organization is to keep records of the results of supplier evaluation and any actions that result from such evaluations.	The purchasing manager establishes and maintains the approved critical vendor list (ACVL), which contains all class A subcontractors (i.e., critical suppliers). Class B and C suppliers are used at the full discretion of the buyers and are recorded separately in the maintenance, repair, and operations (MRO) master list. Supplier evaluations and surveys are recorded in the vendor history files. Specific operational actions taken with regard to suppliers are documented on supplier corrective action reports (SCARs) or receiving inspection reports (RIRs).

We conclude that unless each SHALL of the Standard is addressed clearly within the manual—in the form of quality policy statements—there is a high probability that key sections of the three ISO 9000 pillars (i.e., documentation, implementation, and demonstration of effectiveness) will be trivialized and undermine the integrity of the entire Standard.

5.12 Quality Manual Sequences

5.12.1 Four Possible Quality Manual Sequences

The actual structure of the manual depends on the nature of the enterprise and the manner in which we intend to propagate information within the

QMS. At least four basic configurations for the manual are compliant with the Standard's requirements (as long as the relationship to each section and clause of the Standard is clearly defined by means of either cross-reference charts or references within the text.):

1. Direct sequence based on the Standard's sequence (i.e., Sections 4.0, 5.0, 6.0, 7.0, and 8.0) is compliant. This is also the cut-and-paste method, in that the ISO 9001:1994 manual is edited into an ISO 9001:2000 format.

2. Shewhart cycle sequence (i.e., plan, do, check, act discussed earlier—the direct sequence approximates this sequence) is compliant.

3. Operational cycle sequence (e.g., marketing and sales, engineering, production control, purchasing, receiving, kitting, assembly, test, shipping, customer service—here again the direct sequence, especially Section 7.0, approximates this sequence) is compliant.

4. Another Standard's sequence (e.g., either FDA/CGMP, or EN 46001, which are presently based on the ISO 9001:1994 Standard), or the direct revision of the present ISO 9001:1994 Quality Manual (i.e., keeping the 20 ISO 9001:1994 sections and adding the additional ISO 9001:2000 requirements) is also compliant.

5.12.2 Direct Sequences

The pertinent clauses (SHALLS) of the Standard are located in five sections numbered consecutively from 4.0: Quality Management System to 8.0: Measurement, Analysis, and Improvement. As it was for the 1994-version quality manuals, it is already very common to find manuals configured in this fashion for the 2000 version (i.e., as sections that correspond directly with the numbering system of the Standard). We find this to be true for quality manuals that have been upgraded from the 1994 version to the 2000 version, and for those who have created their initial quality manual for their first ISO certification. We will now describe the methods required in manual creation for both situations.

Manual labels by configuration are as follows:

- An ISO 9001:2000 manual that uses the Standard's numbering system will be termed *manual:2000* (i.e., numbered Sections 4 through 8).
- An ISO 9001:2000 manual that uses the 1994 numbering system will be termed *manual:2000(20)* (i.e., numbered Sections 1 through 20).

- The present ISO 9001:1994 manual will be termed *manual:1994.*

Figure 5.4 is a pictorial view of the several configurations that can occur when an available manual:1994 is upgraded into either the manual:2000 or manual:2000(20) configurations. Notice that in both cases there are a number of additional 2000 requirements that must be addressed to bring the 1994 manual in conformance with the Standard. The difference in effort between the two configurations is the cut-and-paste effort required for the manual:2000, which takes approximately one to two days sitting in front of a computer (this means about two weeks for the average overloaded ISO 9000 management representative). Adding the additional requirements afterwards realistically takes a month of effort and is independent of which configuration is chosen. The toughest sections are 4.0 and 5.0 [22].

The key advantage of the manual:2000 configuration is the ease of auditing the system—because an elaborate cross-reference chart is not required—

Figure 5.4 Configurations for upgraded 1994 manuals (first view).

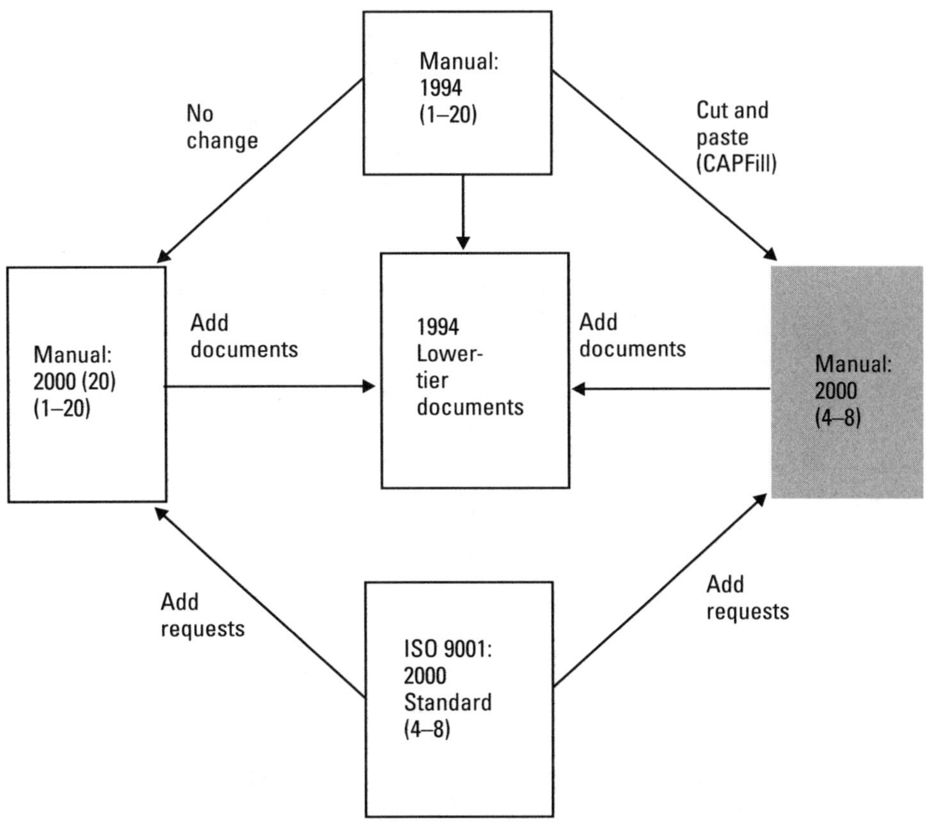

and the use of the more logical operations format of the Standard. Both ways require modifications to the lower tier documents.

I strongly prefer the manual:2000 configuration for the upgrade effort because it is an easier configuration to audit and it directly reflects the process orientation of the Standard.

5.12.2.1 Ground Floor Manual:2000 Creation

In this situation, the organization creates their first manual:2000 based on the Standard's numbering system. This is the path usually traveled for those who have not yet been certified.

The process required to create the initial manual:2000 is as follows:

1. The manual:2000 is formatted into the five directly sequenced sections of the Standard (i.e., Sections 4.0 through 8.0).

2. Quality policy statements are written against every SHALL of the Standard on a 1:1 correspondence with the Standards, numbering system (e.g., Section 4.0 contains all of the requirements for 4.1: General Requirements, 4.2.1: General Documentation Requirements, 4.2.2: Quality Manual, 4.2.3: Control of Documents, and 4.2.4: Control of Records).

3. In the case of regulatory or statutory Standards (e.g. FDA/CGMP 320 or EN46001), be sure to include all of those requirements within the manual:2000. This can be done by either placing the pertinent paragraphs within the four to eight sections or by creating a new section (e.g., "Section 9—Regulatory Affairs," or "Section 10—Security").

We have discovered that the creation of a first manual:2000 depends heavily on whether the consultant has previously written a manual:1994. If everyone is pure of heart, the manual:2000 tends to look very much like the process charts in the Standard. When the consultant has experience on a manual:1994, the manual:2000 tends to reflect the documentation structure of the 1994 version, and there may be a multitude of leftover 1994 typos. Both approaches work just fine!

However, the experienced consultant who uses 1994 documentation structure has little trouble with the trap set in Par. 4.2.1(d) that requires that the QMS contain documents needed by the organization to ensure the effective planning, operation, and control of its processes. Version 1994 requires a wide variety of procedures (i.e., process descriptions) that cover this requirement almost automatically.

5.12.2.2 1994 Upgrade Manual:2000 Creation

In this situation, the organization creates a manual:2000 based on the Standard's numbering system by first cutting and pasting the present manual:1994 into the manual:2000.

Those who have upgraded their 1994 quality manuals soon discover that all of the 1994 quality manual can be cut and pasted into the new format (e.g., 4.5: Document and Data Control:1994 and 4.16: Control of Quality Records: 1994 paste directly into Section 4.0:2000; 4.17: Internal Quality Audits: 1994 and 4.14: Corrective and Preventive Action:1994 paste directly into Section 8.0:2000).

This is true because all of manual:1994 is either just as meaningful in the process of maximizing QMS effectiveness, or, in some cases, is beyond the 2000 version requirements. Don't throw anything away! We will need to enhance the 1994 version but will not want to waste any of it.

5.12.2.3 Cut-and-Paste Technique

The cut-and-paste technique is a relatively painless way to create a substantial part of manual:2000 from what already exists in manual:1994. It is usually done as follows:

1. Manual:2000 is formatted into the five directly sequenced sections of the Standard (i.e., Sections 4.0 through 8.0).

2. Each clause of manual:1994 is cut and pasted into the manual:2000 template, as appropriate, based on the Standard's Appendix B, which describes in detail the correspondence between ISO 9001:2000 and ISO 9001:1994.

3. Each section of manual:2000 is audited against the Standard to determine where manual:2000 created from manual:1994 does not comply with the Standard (e.g., in Section 4.0, there will be significant additional effort over the 1994 quality policy statements needed to respond to 4.1: General Requirements, 4.2.1: General Documentation Requirements, and 4.2.2: Quality Manual, although 4.2.3: Control of Documents and 4.2.4: Control of Records will be in very close conformity with the Standard, including already documented procedures). See Table 5.7.

4. In the case of regulatory or statutory Standards (e.g., FDA/CGMP 320 or EN46001) be sure to cut and paste all of those requirements within the manual:2000. If they were already in manual:1994, and you have fully cut and paste every clause, this issue will automatically have been resolved.

Table 5.7
Required Additional Quality Policy Statements (QPSs) When a Manual:1994 Is Available—Used When Adding Additional ISO 9001:2000 QPSs to a Manual:2000 Formed from Cut-and-Pasted ISO 9001:1994 QPSs or a Manual:1994

ISO 9001:1994 Element	Associated ISO 9001:2000 Element(s)	Additional or Expanded Quality Policy Statements Required To Bring 1994 Requirements in Conformance with the Standard: Description of the Policy Adopted or Methods Used	Potential Impact on Documents
4.1.1: Quality Policy	8.2.1: Customer Satisfaction	Monitor information related to customer perception	Tiers II–IV
		Apply customer perception information as one way to measure QMS performance	
	5.1: Management Commitment	Continually improve QMS effectiveness through the following:	Specifications and reports
		Communicating the importance in meeting customer requirements, including statutory and regulatory requirements	
		Establishing the quality policy	
		Ensuring that quality objectives are established	
		Conducting management review(s)	
		Ensuring resource(s) availability	
	5.3: Quality Policy	Ensure that the quality policy does the following:	Quality policy statement
		Reflects the organization's purpose appropriately	
		Includes a commitment to comply with the QMS' requirements based on the Standard	
		Includes a commitment to continually improve the QMS	
		Provides a framework in which to establish and review quality objectives	
		Is reviewed for continuing suitability	
	5.4.1: Quality Objectives	Ensure that the quality objectives are as follows:	Executive memo
		Established to meet product requirements	
		Established at relevant organizational functions and levels	
		Measurable	
		Consistent with the quality policy	
	5.5.3: Internal communication	Ensure that appropriate communication processes are established	Tier II
		Ensure that communication takes place with regard to QMS effectiveness	
4.1.2.1: Responsibility and Authority	5.5.1: Responsibility and Authority	Ensure that responsibilities and authorities are communicated within the organization	Executive memo

Table 5.7 (continued)

ISO 9001:1994 Element	Associated ISO 9001:2000 Element(s)	Additional or Expanded Quality Policy Statements Required To Bring 1994 Requirements in Conformance with the Standard: Description of the Policy Adopted or Methods Used	Potential Impact on Documents
4.1.2.2: Resources	6.1: Provision of Resources	Ensure that the resources needed to continually improve QMS effectiveness is determined and provided	Budgets and business plans
		Ensure that the resources needed to enhance customer satisfaction is determined and provided	
	6.2.1: General	Ensure that personnel performing work affecting product quality are competent on the basis of appropriate education, training, skills, and experience	Tier IV
4.1.2.3: Management Representative	5.5.2: Management Representative	Ensure that the needed processes are established, implemented, and maintained	Tier II posters
		Report to top management on any need for QMS improvement	
		Ensure the promotion of awareness of customer requirements throughout the organization	
4.1.3: Management Review	5.6.1: General Management Review	Review the QMS at planned intervals to ensure its adequacy	Reports
		Assess opportunities for improvement	
		Assess the need for QMS changes including changes to the quality policy and quality objectives	
	5.6.2: Review Input	Include the following management review inputs:	Reports
		Results of audits	
		Customer feedback	
		Process performance and product conformity	
		Corrective and preventive action (CAPA) status	
		Follow-up actions from previous reviews	
		Changes that could affect the QMS	
		Recommendations for improvement	
	5.6.3: Review Output	Include management review outputs with regard to decisions and actions to do the following:	Reports
		Improve the QMS effectiveness and its processes	
		Improve product against customer requirements	
		Provide needed resources	

Table 5.7 (continued)

ISO 9001:1994 Element	Associated ISO 9001:2000 Element(s)	Additional or Expanded Quality Policy Statements Required To Bring 1994 Requirements in Conformance with the Standard: Description of the Policy Adopted or Methods Used	Potential Impact on Documents
	8.5.1: Continual Improvement	Continually improve QMS effectiveness via use of the following:	Reports
		Quality policy	
		Quality objectives	
		Audit results	
		Analysis of data	
		CAPA	
		Management review	
4.2.1: General	4.1: General Requirements	Implement the QMS	Tier II
		Continually improve its effectiveness	
		Identify the processes needed for the QMS	
		Identify how the processes are applied within the organization	
		Determine the sequence and interaction of the processes	
		Determine criteria and methods needed to ensure process operation and control is effective	
		Ensure the availability of resources and information to support process operation and monitoring	
		Implement actions necessary to achieve planned process results	
		Implement actions necessary to achieve continual process improvement	
		Manage the organization in conformance with the Standard	
		Control outsourced processes	
4.2.2: Quality System Procedures	4.2.1: General	Include documented statements of a quality policy	Tiers II–IV
		Include documented statements of quality objectives	
		Include documents needed to ensure effective planning in the operation and control of the processes	
4.2.3: Quality Planning	5.4.2: QMS Planning	Ensure that quality objectives are planned	Business plans
		Ensure that QMS integrity is maintained when QMS changes are planned and implemented	

Table 5.7 (continued)

ISO 9001:1994 Element	Associated ISO 9001:2000 Element(s)	Additional or Expanded Quality Policy Statements Required To Bring 1994 Requirements in Conformance with the Standard: Description of the Policy Adopted or Methods Used	Potential Impact on Documents
	7.1: Planning of Product Realization	Plan and develop QMS consistent product realization processes Determine product quality objectives and requirements Determine the need for processes Determine the need to verify, validate, and monitor product performance against product acceptance criteria	Business plans and quality/control plans
4.3.2: Review	5.2: Customer Focus	Determine customer requirements Enhance customer satisfaction Monitor information related to customer perception Apply customer perception information as one way to measure QMS performance	Tier II
	7.2.1: Determination of Requirements Related to the Product	Determine specified customer requirements related to post-delivery activities Determine nonstated customer requirements but also necessary for the specified and/or intended use Determine statutory and regulatory requirements Determine any additional requirements the organization feels is required	Tier II
	7.2.2: Review of Requirements Related to the Product	Maintain actions arising from the requirements review Note: include internet sales reviews, if pertinent	Tier II
	7.2.3: Customer Communication	Determine and implement effective arrangements for customer communication with regard to the following: Product information Customer feedback, including customer complaints	Tier II
4.4.2: Design and Development Planning	7.3.1: Design and Development Planning	Review, verify, and validate each design and development (D&D) stage Determine D&D authorities	Tier II
4.4.3: Organizational and Technical Interfaces	7.3.1: Design and Development Planning	Ensure effective interface communication Ensure clear assignment responsibility	Tier IV

Table 5.7 (continued)

ISO 9001:1994 Element	Associated ISO 9001:2000 Element(s)	Additional or Expanded Quality Policy Statements Required To Bring 1994 Requirements in Conformance with the Standard: Description of the Policy Adopted or Methods Used	Potential Impact on Documents
4.4.4: Design Input	7.2.1: Determination of Requirements Related to the Product	Determine specified customer requirements related to post-delivery activities Determine nonstated customer requirements but also necessary for the specified and/or intended use Determine any additional requirements the organization feels is required	Tier II
	7.3.2: Design and Development Inputs	Maintain records Include functional and performance requirements Include, where applicable, information derived from previous similar designs Include other requirements essential for D&D	Tier II
4.4.5: Design Output	7.3.3: Design and Development Outputs	Provide appropriate information for purchasing, production, and service provision	Tier II
4.4.6: Design Review	7.3.4: Design and Development Review	Evaluate the ability of the results to fulfill requirements Identify any problems and propose necessary actions	Tiers II and IV
4.4.7: Design Verification	7.3.5: Design and Development Verification	Maintain a record of necessary actions that result from verification	Tier IV
4.4.8: Design Validation	7.3.6: Design and Development Validation	Validate product performance prior to delivery or implementation, wherever practicable Maintain records and any necessary actions that result	Tier IV
4.4.9: Design Changes	7.3.7: Control of Design and Development Changes	To maintain records of D&D changes with necessary actions included Verify and validate D&D changes, as appropriate Evaluate the effect of the changes on constituent parts Evaluate the effect of the changes on product already delivered	Tier II
4.5.2: Document Data Approval Issue	7.2.3: Control of Documents	To reapprove documents Ensure legibility and ready identification Ensure unintended use of obsolete documents retained for any purpose	Tier II

Table 5.7 (continued)

ISO 9001:1994 Element	Associated ISO 9001:2000 Element(s)	Additional or Expanded Quality Policy Statements Required To Bring 1994 Requirements in Conformance with the Standard: Description of the Policy Adopted or Methods Used	Potential Impact on Documents
4.6.2: Evaluation of Subcontractors	7.4.1: Purchasing Process	Ensure that subcontractors are reevaluated Maintain records of necessary actions that result from subcontractor evaluation	Tiers II and IV
4.6.4: Verification of Purchased Product	7.4.3: Verification of Purchased Product	Establish and implement inspection or other activities required to ensure that purchased product meets specifications Ensure that when your customer performs source inspection we have stated the verification arrangements and method of product release in the purchasing information	Tier II
4.7: Control of Customer Supplied Product	7.5.4: Customer Property	Identify, protect, and safeguard customer property Note: include intellectual property in this control	Tier II
4.8: Product Identification and Traceability	7.5.3: Identification and Traceability	Identify product status with respect to monitoring and measurement Control the unique identification of the product Note: include configuration management within this control, if appropriate	Tiers III and IV
4.9: Process Control	6.3: Infrastructure	Determine, provide, and maintain an infrastructure designed to achieve conformity to product requirements, consider the following: Buildings, workspaces, and associated utilities Process equipment (both hardware and software) Supporting services such as transport or communication	Tiers II–IV
	6.4: Work Environment	Determine and manage the work environment required to achieve conformity to product requirements	Tiers II–IV
	7.5.1: Control of Production and Service Provision	Include controlled conditions, as applicable, for the following: Information that describes product characteristics Work instructions, as necessary Monitoring and measuring devices Implementation of monitoring and measurement Implementation of release, delivery and postdelivery activities	Tiers II–IV

Table 5.7 (continued)

ISO 9001:1994 Element	Associated ISO 9001:2000 Element(s)	Additional or Expanded Quality Policy Statements Required To Bring 1994 Requirements in Conformance with the Standard: Description of the Policy Adopted or Methods Used	Potential Impact on Documents
	7.5.2: Validation of Processes for Production and Service Provision	Validate the ability of special processes to achieve planned results Revalidate special processes Show approval of equipment	Tiers III and IV
4.10.2–4.10.5: Inspection and Testing	8.2.4: Monitoring and Measurement of Product	Monitor and measure the characteristics of the product Indicate in the records the person(s) authorizing release of the product Release product and service delivery until planned arrangements are completed or otherwise approved by a relevant authority and, where applicable, by the customer	Tiers III and IV
4.10.1: General	7.1: Planning of Product Realization	Determine the need to verify, validate, and monitor product performance against product acceptance criteria	Tier II
4.10.1: General	8.1: General	Plan and implement the monitoring, measurement, analysis and improvement processes needed to continually improve the effectiveness of the QMS	Tier II
4.10.2: Receiving Identification and Traceability	7.4.3: Verification of Purchased Product	State intended verification arrangements and methods of product release in the purchasing information when the organization or its customer intends to perform source inspection	Tiers III and IV
4.10.5: Identification and Traceability Records	7.5.3: Identification and Traceability	Identify product status with respect to monitoring and measurement requirements	Tier IV
4.11.1– 4.11.2: Control of Inspection, Measuring, and Test Equipment	7.6: Control of Monitoring and Measuring Devices	Determine the monitoring and measurement to be undertaken Establish processes for this purpose Calibrate or verify at specified intervals or prior to use Adjust or readjust devices as necessary Protect devices from damage and deterioration during maintenance Ensure that when equipment that was used out of calibration is found, the organization takes appropriate action on the equipment and any product affected Ensure that the ability of computer software to satisfy the intended application is confirmed. Confirmation is to occur prior to initial use and reconfirmed as necessary	Tiers II–IV

Table 5.7 (continued)

ISO 9001:1994 Element	Associated ISO 9001:2000 Element(s)	Additional or Expanded Quality Policy Statements Required To Bring 1994 Requirements in Conformance with the Standard: Description of the Policy Adopted or Methods Used	Potential Impact on Documents
4.12: Identification and Traceability Status	7.5.3: Identification and Traceability	Identify product status with respect to monitoring and measurement requirements	Tier IV
4.13: Control of Nonconforming Product	8.3: Control of Nonconforming Product	Ensure that release or acceptance under concession is by a relevant authority, and, where applicable, by the customer Maintain records of the nature of the nonconformities and any subsequent actions taken, including concessions obtained	Tiers II–IV
4.14: CAPA	8.5.2: Corrective Action	Ensure that actions are taken to eliminate the cause of nonconformities in order to prevent recurrence	Tier II
4.14: CAPA	8.5.3: Preventive Action	Ensure that actions are taken to eliminate the causes of potential nonconformities in order to prevent their occurrence Ensure that preventive actions are appropriate to the effects of the potential problems Maintain records of the results of actions taken	Tiers II and IV
4.15: Handling, Storage, Preservation, Packaging, and Delivery	7.5.5: Preservation of Product	Apply preservation to the constituent parts of a product	Tiers III and IV
4.15.6: Delivery	7.5.1: Control of Production and Service Provision	Ensure that postdelivery activities are controlled	Tiers II and IV
4.17: Internal Quality Audits	8.2.2: Internal Audit	Define the audit criteria, scope, frequency, and methods Select auditors and conduct audits such that objectivity and impartiality are ensured in the audit process Ensure that auditors do not audit their own work	Tiers II and IV
	8.2.3: Monitoring and Measurement of Processes	Ensure that the methods used to monitor and measure the QMS processes have the ability to achieve planned results Ensure that when planned results are not achieved correction and corrective action is taken as appropriate	Tiers II and IV

Table 5.7 (continued)

ISO 9001:1994 Element	Associated ISO 9001:2000 Element(s)	Additional or Expanded Quality Policy Statements Required To Bring 1994 Requirements in Conformance with the Standard: Description of the Policy Adopted or Methods Used	Potential Impact on Documents
4.18: Training	6.2.1: General	Demonstrate that personnel performing work affecting product quality are competent on the basis of appropriate education, training, skills, and experience	Tiers III and IV
	6.2.2: Competence, Awareness, and Training	Determine the necessary competence for personnel performing work affecting product quality	Tiers II and IV
		Provide training or take other steps to satisfy training needs	
		Evaluate the effectiveness of the training	
		Ensure that its personnel are aware of the relevance and importance of their activities and how they contribute to the achievement of the quality objectives	
4.19: Servicing	7.5.1: Control of Production and Service Provision	Ensure the availability of the following:	Tiers II–IV
		Product characteristic information	
		Work instructions, as necessary	
		Suitable equipment	
		Monitoring and measuring devices	
4.20.1: Identification of Need	8.1: General	Determine the extent to which statistical techniques are to be used	Tier II
4.20.1: Identification of Need; 4.20.2: Procedures	8.2.3: Monitoring and Measurement of Processes	Apply suitable methods to do the following:	Tiers II and IV
		For monitoring and, where applicable, measurement of the QMS processes	
		To demonstrate the ability of the processes to achieve planned results	
	8.2.4: Monitoring and Measurement of Product	Monitor and measure (M&M) the characteristics of the product to verify conformance to requirements	Tiers II and IV
		Carry out M&M of characteristics at appropriate stages of product realization	
		Maintain evidence of conformity with the acceptance criteria	

Table 5.7 (continued)

ISO 9001:1994 Element	Associated ISO 9001:2000 Element(s)	Additional or Expanded Quality Policy Statements Required To Bring 1994 Requirements in Conformance with the Standard: Description of the Policy Adopted or Methods Used	Potential Impact on Documents
	8.4: Analysis of Data	Determine, collect, and analyze appropriate data to do the following:	Tiers II and IV
		Demonstrate the suitability and effectiveness of the QMS	
		Evaluate where continual improvement of QMS effectiveness can be made	
		Include data generated as a result of M&M and from other relevant sources	
		Analyze data to provide information relating to do the following:	
		Customer satisfaction	
		Conformity to product requirements	
		Characteristics and trends of process and products including opportunities for preventive action	
		Suppliers	

Alert: The cut-and-paste transformation of the manual:1994 into manual:2000 does have some difficulties because the 1994 clauses are mixed about within the 2000 sections in some novel ways (e.g., the 4.10: Inspection & Test:1994 clauses are distributed in Section 7: Product Realization:2000 and Section 8: Measurement, Analysis, & Improvement:2000. To make things a bit more complex, pertinent 4.10 clauses appear in the five new Clauses 7.1, 7.4.3, 7.5.3, 8.1, and 8.2.4. However, the Standard has worked diligently to make the cut-and-paste operation as smooth as possible via a cross-reference chart between the clauses (refer to pp. 19–22 of the Standard). In particular, if you follow Table B.1—Correspondence between ISO 9001:1994 and ISO 9001:2000, on pp. 19 and 20 of the Standard, your cut-and-pasting exercise should take less than 2 days of work.

5. Fill in the additional quality policy statements that are required above and beyond the 1994 requirements. This process is reciprocal to the fourth technique in which we keep the 1994 version numbering system and add the additional 2000 requirements. See Figure 5.5.

Table 5.8 is a cut-and-paste at-a-glance chart that provides a simplified overview of the issues (i.e., the clauses that need to be broken up into their appropriate quality policy statements and placed within the Standard's five sections). For example, the five clauses of 4.1: Management Responsibility:1994 end up in nine different clauses and in three different sections in the 2000 version. By contrast, 4.4: Design Control:1994 is found only in Clause 7.3: Design and Development:2000.

Figure 5.5 Upgrading a 1994 manual to a 2000 version manual.

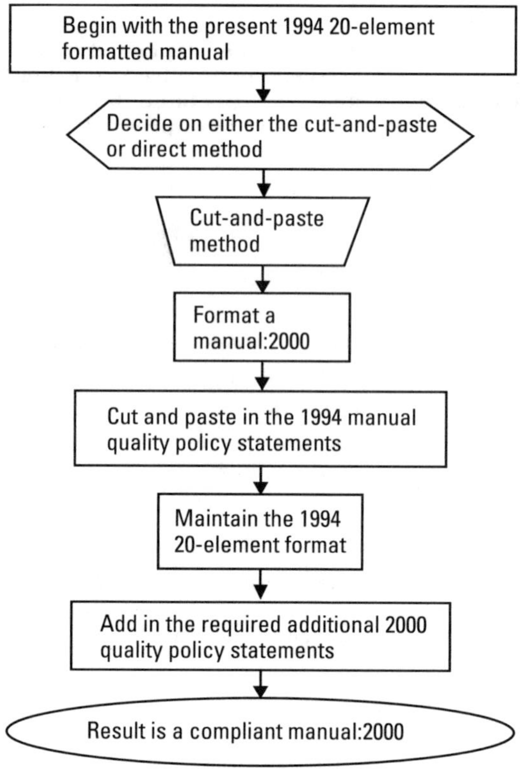

Table 5.8
ISO 9001:1994 Cut-and-Paste-at-a-Glance Chart (Use When a 1994 Quality Manual Already Exists and a 2000 Manual Is Formatted)

1994 Clauses	ISO 9001:1994 Titles[1]	Manual:2000 Sections[2]				
		4.0: QMS	**5.0: Management Responsibility**	**6.0: Resource Management**	**7.0: Product Realization**	**8.0: Measurement, Analysis, and Improvements**
4.1	Management Responsibility		5.1, 5.3, 5.4.1, 5.5.1, 5.5.2, 5.5.3 (new), 5.6.1	6.1, 6.2.1		8.5.1
4.2	Quality System	4.1, 4.2.1, 4.2.2	5.4.2 (Planning)		7.1 (Planning)	
4.3	Contract Review		5.2 (Customer Focus)		7.2.1, 7.2.2, 7.2.3	8.2.1 (new)

Table 5.8 (continued)

1994 Clauses	ISO 9001:1994 Titles[1]	Manual:2000 Sections[2]				
		4.0: QMS	5.0: Management Responsibility	6.0: Resource Management	7.0: Product Realization	8.0: Measurement, Analysis, and Improvements
4.4	Design Control				7.2.1, 7.3	
4.5	Document and Data Control	4.2.3				
4.6	Purchasing				7.4.1, 7.4.2, 7.4.3	
4.7	Control of Customer Supplied Product				7.5.4	
4.8	Product Identification and Traceability				7.5.3	
4.9	Process Control			6.3, 6.4	7.5.1, 7.5.2	
4.10	Inspection and Testing				7.1, 7.4.3, 7.5.3	8.1, 8.2.4
4.11	Control of Inspection, Measuring, and Test Equipment				7.6	
4.12	Inspection and Test Status				7.5.3	
4.13	Control of Nonconforming Product					8.3
4.14	Corrective and Preventive Action					8.5.2, 8.5.3
4.15	Handling, Storage, Packaging, Preservation and Delivery				7.5.1, 7.5.5	
4.16	Control of Quality Records	4.2.4				
4.17	Internal Quality Audits					8.2.2, 8.2.3
4.18	Training			6.2.2		
4.19	Servicing				7.5.1	
4.20	Statistical Techniques					8.1, 8.2.3, 8.2.4, 8.4

Source: [23]. Alert: Neither 5.5.3 or 8.2.1 is assigned in the Standard's cross references.

1. Cut and paste these clauses into the appropriate manual:2000 sections.

2. Paste appropriate1994 clauses into these sections.

5.12.2.4 The Fill-In Process

After you have cut and pasted the 1994 clauses into manual:2000, the hard part starts because we must now fill in the quality policy statements that are dictated by the Standard above and beyond the 1994 requirements. Table 5.7 summarized to a high degree of exactitude what must be added to manual:2000 to make it conform in detail with the Standard. Reciprocally, it is also the *differences* between ISO 9001:1994 and the Standard that must be added to manual:2000(20) to make it conform with the Standard.

Example of Tier I Response Table 5.7 includes a final column that addresses the "Potential Impact on Documentation" caused by the additional 2000 requirements. For example, Clause 8.2.1 of the Standard adds the need to describe a policy with regard to monitoring information as it relates to a customer's perception of the company's service. In this case, the quality policy statement in manual:2000 might sound something like this:

> *8.2.1 Customer Perception of Excellent*: Customer perception of Excellent's performance is monitored and measured rigorously by means of weekly sample interviews with key customers by the national sales manager; monthly reviews by quality assurance with the executive committee on customer complaints and customer commendations; and monthly customer service reports on all aspects of service performance. Metrics are established for each of these activities (refer to Doc. #MS3-08 entitled, "Marketing and& Sales Customer Metrics").

Potential Tier II Impact This response would probably require a modification to, let's suppose, the present tier II document #MS2-001 entitled, "Marketing and Sales Procedures," the creation of the new tier III document #MS3-08, and the creation of a new tier IV form #MS4-009 entitled "Marketing and Sales Metric Performance."

As indicated in Table 5.7, there are about 170 possible quality policy statement enhancements needed to bring the 1994 requirements in conformance with the Standard. As noted, these modifications can ripple through the lower level documents and have the potential to cause multiple revisions throughout the QMS documentation.

For those readers who have been blessed with aggressive auditors—who have constantly pushed the envelope of your quality system to consider all of your core competencies—the amount of revision will be reasonable. For others who are not so fortunate, the amount of revision could be daunting. Allow yourself plenty of time to work on your upgrades.

Many clients are giving themselves a year or so and are timing their upgrades to a surveillance assessment (e.g., carry out the document review three months before the surveillance, the preassessment two months before the surveillance, and the upgrade assessment at the same time as the surveillance audit). Unfortunately, many upgraders are waiting until the last minute to upgrade, which could cause a serious scheduling problem among registrars [24].

5.12.2.5 Assessment Implications for Directly Sequenced Manuals

ISO 9001:2000 auditors commonly assume the existence of a directly sequenced manual [25]. Furthermore, they generally assume a one-to-one correspondence with each clause in the element. The form of the certification assessment generally takes on the form shown in Table 5.9 [26].

Notice the way that certain processes precede other processes in the audit flow:

- Top management process precedes QMS structure as a way to achieve an umbrella perspective of the management style and motivation before an audit of detailed operations;
- Marketing and sales process precedes design and development process because it is the marketing requirements documentation that drives design engineering;
- Customer property process precedes service provision process for customer in the application of that clause to service functions because, for example, returned goods often belong to the customer during the repair process.

Such audit patterns are designed to capture the concomitant relationships of one Standard's element to another. In addition, registrar assessors and other third-party or second-party auditors tend to gravitate towards this type of element alignment because it is an efficient way to do audits against very stringent time constraints.

In a certification audit, the assessor must keep moving along and cover all of the elements within the time frames noted in the far left column. There is very little opportunity to go back and check out an observation later on in the audit. In fact, it is best to tell the auditee(s) exactly when the audit is frozen and not drive them crazy with sudden appearances at the last minute to decide on a nonconformance report. My policy has always been to tell the client, for example, "We will end the audit at 4:00 P.M.!" and make that time within plus or minus one minute. I seldom miss.

Table 5.9
ISO 9001:2000 Certification Assessment for Excellent, Inc.

Time	Activities	Auditors/Location		Initial Auditee(s)
Day One				
8:15 A.M.	Arrive on site	Team	Corporate conference room	Core group
8:30	Opening meeting chaired by the lead auditor			
9:00	Tour of the plant	Team	Quick walk of facility	Management representative
9:30	Top-management process: 5.1, 5.3, 5.4.1, 5.5.1, 5.5.2, 5.5.3, 5.6.1, 6.1, 6.2.1, 8.5.1	Team	Conference room	Core-group management
11:00	QMS structure: 4.1, 4.2.1, 4.2.2, 5.4.2, 7.1	Lead	Quality-assurance office	Management representative
11:00	Marketing and sales process: 5.2, 7.2.1,2,3, 8.2.1	Auditor 2	Marketing and sales	V.P. of marketing and sales
Noon	Lunch	Team	Fast food	Management representative
1:00	Internal audit process: 8.2.2,&3	Lead	Quality-assurance office	Quality-assurance manager
1:00	Design and development process: 7.2.1(Reg), 7.3	Auditor 2	Engineering area	V.P. of engineering
2:00	Corrective and preventive action process: 8.5.2,3	Lead	Quality-assurance office	Quality-assurance manager
3:00	Analytical process: 8.1, 8.2.3, 8.2.4, 8.4	Lead	Quality-assurance office	Quality-assurance manager
3:00	Purchasing process: 7.4	Auditor 2	Purchasing	Purchasing manager
3:30	Prepare for daily review	Team	Prep office	
4:00	Daily review—chaired by lead auditor	Team	Conference room	Core group
4:30	Exit site	Team		
Day Two				
8:30 A.M.	Review of previous day's findings	Team	Conference room	Core group
9:00	Document control process: 4.2.3	Lead	DC office	DC coordniator
9:00	Production control process: 6.3, 6.4, 7.5.1, 2 plus control of nonconforming product: 8.3	Auditor 2	Manufactuting areas	V.P. of operations
10:00	Training process: 6.2.2	Lead	HR office	HR manager
10:30	Identification and traceability process: 7.5.3	Auditor 2	Manufacuring areas	V.P. of operations

Table 5.9 (continued)

Time	Activities	Auditors/Location		Initial Auditee(s)
11:00	Control of monitoring and measuring devices process: 7.6	Lead	Metrology	Calibration supervisor
11:00	Monitoring and measuring of product process: 7.1, 7.4.3, 7.5.3, 8.1, 8.2.4	Auditor 2	Manufactuting and quality-assurance areas	V.P. of operations. and quality-assurance manager
11:30	Preservation of product processes: 7.5.1, 7.5.5	Auditor 2	Manufactuting areas	V.P. of operations
12:00	Lunch	Team	Fast food	Management representative
1:00	Control of records process: 4.2.4	Lead	Quality-assurance office	Quality-assurance manager
1:00	Customer property process: 7.5.4 plus service provision process: 7.5.1	Auditor 2	Customer service	Customer-service manager
2:00	Auditors prepare for closing meeting	Team	Prep office	
4:00	Closing meeting chaired by lead auditor	Team	Conference room	Core group
4:30	Auditors exit site	Team		

Table 5.10 summarizes what I have found to be the most difficult implementation issues for those either upgrading or initializing a QMS based on the Standard. Unfortunately, the changeover to the Standard is moving slowly, and this analysis is based on only 10 ISO 9001:2000 certifications. However, inputs from other assessors point to the same issues. The setting of objectives and implementation of audits are the two most difficult tasks.

5.12.2.6 Checklist

If the direct-sequence method is chosen, a convenient checklist can be generated to track manual progress as part of an extensive quality manual desk audit. An example of such a checklist is shown in Table 5.11.

5.12.2.7 Readiness Concept

As a rule, when a section is rated at 90% or higher, it is ready for the initial systems assessment (i.e., certification assessment sometimes called the "A-1"). That means that, alternately, we are ready to do the preassessment (PA-1) to fine tune the system prior to the certification audit.

Some companies choose to go directly to the certification audit. However, my experience with over 100 certifications in the United States and Canada is that a preassessment is a very good idea [27]. All of the open nonconformance

Table 5.10
ISO 9001:2000 Most Difficult Critical Issues (in Order of Difficulty)

Element(s)	Title of Clause	Degree of Difficulty
5.4.1 + 7.1(a)	Quality Objectives (ability to define objectives, metrics, and goals)	Most
	Planning of Product Realization (quality objectives for the product)	Most
8.2.2	Internal Audit (coverage of all processes against appropriate SHALLS)	Most
4.1	General Requirements (creation of appropriate process documents)	High
4.2.1(d)	General Documentation Requirements (other documents other than procedures)	High
4.2.3	Control of Documents (methods of control and use of electronic media)	High
8.5.3	Preventive Action (interpretation of what constitutes a preventive action)	High
8.5.2(a) + 8.2.1	Customer Complaints (completeness of coverage and closure/follow up)	High
	Customer Satisfaction (or dissatisfaction)	High
8.5.2	Corrective Action (closure—e.g., follow-up actions)	Moderate
4.2.4	Control of Records (what is a record and how long need they be kept?)	Moderate
5.5.2(c)	Management Representative (ensuring promotion of quality awareness)	Moderate

reports (NCRs) should be rigorously responded to and considered closed by the organization prior to the PA-1. There must be no majors anywhere in the system at that point, so that it will be possible to truly judge the status of the QMS with all of the documentation in place. This rule also holds after the PA-1 and prior to the certification assessment so that it will be possible to judge the effectiveness of the QMS with all the documentation and implementation in place and operational. Most companies are able to accomplish this task with a good deal of hard work. This means that all of management must be part of this commitment to excellence.

The comments in Table 5.11 indicate what type of specific action item is required to complete the task and whether a specific NCR has been written against the section during an internal quality audit. There were 17 NCRs written against the system with only two closed at the moment. There are four open majors.

Table 5.11
Excellent's ISO 9001:2000 Quality Manual Status—Direct Sequence Manual

Activities	Percentage Found in Text Actual/ *Plan*	Nonconformances Issued	Comments (End of Week 12)
ISO 9001:2000 Clauses (1994)			
Quality manual cover page	95/*90*		Ready for PA-1
Table of contents	95/*90*		Ready for PA-1
Section 1.0: History of excellent	95/*90*		Ready for PA-1
Section 2.0: Scope of the certification	80/*90*		Firm up wording
Section 3.0: Quality policy statement	80/*90*		Needs posting
Section 4.0: Quality management system	80/*90*		Clean up NCRs
4.1 General requirements-QMS *(4.2.1)*	85	NCR001	Minor: open
4.2.1 General document requirements *(4.2.2)*	85	NCR002	Minor: open
4.2.2 Quality manual *(4.2.1)*	80		Need control stamp
4.2.3 Control of documents *(4.5)*	75	NCR003	Major: open, distribute?
4.2.4 Control of records *(4.16)*	75	NCR004	Minor: open, master?
Section 5.0: Management Responsibility	90/*90*		
5.1 Management commitment *(4.1.1)*	90		Ready for PA-1
5.2 Customer focus *(4.3.2)*	90		Ready for PA-1
5.3 Quality policy *(4.1.1)*	90		Get posters up
5.4.1 Quality objectives *(4.1.1)*	85	NCR005	Minor: open, goals
5.4.2 QMS Planning *(4.2.3)*	85		Rewrite late
5.5.1 Responsibility and authority *(4.1.2.1)*	90		Ready for PA-1
5.5.2 Management representative *(4.1.2.3)*	95		Ready for PA-1
5.5.3 Internal communication *(new)*	85	NCR006	Minor: closed
5.6 Management review *(4.1.3)*	90	NCR007	Minor: closed, report?
Section 6.0: Resource management	90/*90*		
6.1 Provision of resources *(4.1.2.2)*	95		Ready for PA-1
6.2 Human resources *(4.18, 4.1.2.2)*	95	NCR008	Minor: closed, great job
6.3 Infrastructure *(4.9)*	90		Ready for PA-1
6.4 Work Environment *(4.9)*	80		Screen room issue
Section 7.0: Product Realization	84/*90*		
7.1 Planning of Product Realization *(4.2.3, 4.10.1)*	90		Ready for PA-1
7.2.1 Det. of Rqts. related to Product *(4.3.2, 4.4.4)*	90		Ready for PA-1

Table 5.11 (continued)

Activities	**Percentage Found in Text Actual/ *Plan***	**Nonconformances Issued**	**Comments (End of Week 12)**
7.2.2 Review of Requests related to product *(4.3)*	85	NCR009	Minor: open, engineering?
7.2.3 Customer communication *(4.3)*	85		Rewrite late
7.3 Design and development *(4.4)*	80	NCR010	Minor: open, beta tests
7.4 Purchasing *(4.6 w/4.10.2 and 7.4.3)*	80	NCR011	Minor: open, PM
7.5.1 Control of production and service provision *(4.9, 4.15.6, 4.19)*	75	NCR012	Minor: open, D/B due
7.5.2 Val. of processes production and service provision *(4.9)*	95		Ready for PA-1
7.5.3 Identification and traceability *(4.8, 4.10.5, 4.12)*	80		Receiving W/I issue
7.5.4 Customer property *(4.7)*	85		Need better example
7.5.5 Preservation of product *(4.15)*	75	NCR013	Minor: open, ESD issue
7.6 Control monitoring and measuring devices *(4.11)*	85		In-house cal. late
Section 8.0: Measurement, analysis and improvement	78/*90*		
8.1General MA&I *(4.10.1, 4.20.1,2)*	75		Clarity an issue
8.2.1 Customer satisfaction *(new)*	80		Dissatisfaction?
8.2.2 Internal audit *(4.17)*	70	NCR014	Major: open, incomplete
8.2.3 Monitoring and measurement of processes *(4.17,4.20.1,2)*	75		Process definition?
8.2.4 Monitoring and measurement of product *(4.10 and 4.20 w/o procedures)*	85		In-process issue
8.3 Control of nonconforming product *(4.13)*	95	NCR015	Minor: open, scrap?
8.4 Analysis of data *(4.20.1,2)*	80		Where are the charts?
8.5.1 Continual improvement *(4.1.3)*	80		Same as above
8.5.2 Corrective Action *(4.14.1, 4.14.2)*	70	NCR016	Major: open, closure
8.5.3 Preventive Action *(4.14.1, 4.14.3)*	65	NCR017	Major: open, teams?

5.12.3 Shewhart Sequence

The Shewhart cycle of continuous improvement (i.e., plan-do-check-act) can also be used to configure the manual, especially as the Standard has attempted to follow this pattern to a high degree. Table 5.12 provides some idea on how the five sections of the Standard are distributed across the Shewhart cycle.

Table 5.12
Approximate Shewhart Cycle for ISO 9001:2000

ISO 9001:2000 Sections	ISO 9001:2000 Elements	Plan	Do	Check	Act
4.0: Quality Management System (QMS)	4.1: General Requirements		4.1(a), (b), (c), (d)	4.1(e)	4.1(f)
	4.2.1: Documentation Requirements—General	4.2.1(a)	The rest		
	4.2.2: Quality Manual	4.2.2(a)	The rest		
	4.2.3: Control of Documents		All		
	4.2.4: Control of Records		All		
5.0: Management Responsibility	5.1: Management Commitment		All		
	5.2: Customer Focus		All		
	5.3: Quality Policy	Rest		5.3(e)	
	5.4: Planning; Quality Objectives; QMS Planning	5.4.2	5.4.1		
	5.5: Responsibility, Authority and Communication		All		
	5.6: Requirements—Management Review			All	
6.0: Resource Management	6.1: Provision of Resources		All		
	6.2: HR—Competence, Awareness, Training		The rest	6.2.2(c)	
	6.3: Infrastructure		All		
	6.4: Work Environment		All		
7.0: Product Realization	7.1: Planning of Product Realization	All			
	7.2: Customer Related Processes		All		
	7.3: Design and Development	7.3.1	7.3.2,3	7.3.5,6	7.3.4,7
	7.4: Purchasing		All		
	7.5: Production and Service Provision	7.5.1	The rest	7.5.2	
	7.6: Control of Monitoring and Measuring Devices		The rest	7.6(a)	
8.0: Measurement Analysis and Improvement	8.1: General	Split	Split		
	8.2: Monitoring and Measurement	Audits	8.2.3,4	8.2.1,2	8.2.3
	8.3: Control of Nonconforming Product		Split	Split	Split
	8.4: Analysis of Data		Split	Split	
	8.5: Improvement		8.5.1		8.5.2,3

The rest = the remaining clauses.

As illustrated in Table 5.12, the clauses of the Standard can be organized into explicit plan-do-check-act Shewhart Cycle categories (refer to Chapter 3). A manual structured in this fashion would need a cross-reference chart to indicate exactly how the reader was to locate the specific Standard's clauses for either reference or audit purposes.

This approach is perhaps the most logical sequence to follow but it does have a number of problem areas (e.g., design and development and monitoring and measurement), where a split of the clauses might not be the most efficient way to discuss those processes.

We have yet to observe this method used in either a manual:1994 or manual:2000, but I have incorporated the concept into a combined TQM/ISO program [28]. In that approach, the relationship between the Standard's elements and the Shewhart Cycle varies with the TQM concept. Table 5.13 demonstrates this method. The original certification was based on the 1994 version and we have transformed the numbering system to meet the Standard's nomenclature as part of an anticipated upgrade requirement [29].

For example, in the plan cycle, the TQM model includes goals, marketing, estimating, and supplemental control.

A directly integrated TQM/ISO program is designed to gain the full advantage of both quality management concepts. We will cover this topic no further because there are many books available on TQM [30].

5.12.4 Operational Sequence

Another manual layout method is to configure the manual in terms of the actual organization's manufacturing or service processes. We have seen several attempts made at this approach in the 1994 version, but they were eventually rejected due to the correlation difficulty with the Standard's sections. However, as illustrated in Figure 5.6, the Standard's five sections more readily lend themselves to an operational flow [31].

We have applied this approach to the 1994 version by means of a tier II, manufacturing process manual. This type of document is essentially a quality plan because it begins with marketing and sales and flows through to after sales servicing. Its use permits the reader to more readily sense the operational flow as the quality policy manual is read. It's application to the 2000 version would be even more appropriate.

The manufacturing process manual is generally created with flow charts and supplementary text. When flow charts are used, the entire manufacturing process can be posted in key areas of the facility and their presence is an impressive display for visitors.

Value Chain As mentioned previously, Figure 5.6 demonstrates this operational flow inherent in the Standard, as indicated by the central flow of activities from marketing and sales to service, install, and repair. This intrinsic operational flow is supported by a set of executive functions (e.g., management responsibility

Table 5.13
Quality Manual Contents for a Construction Company (Shewhart Cycle Example)

***Plan* Cycle—Goals/Marketing and Sales/Estimating and Supplemental Control**
Section 5.0: Management Responsibility (all clauses)
Section 4.0: Quality Management System (all clauses)
Clause 7.1: Planning of Product Realization
Clauses 7.2.1: Determination of Requirements Related to Product, 7.2.2: Review of Requirements Related to Product, 7.2.3: Customer Communication, 8.2.1: Customer Satisfaction
Clause 7.3: Design and Development
Clause 8.5.1: Continual Improvement
***Do* Cycle—Purchasing and Project Coordination/Manufacturing**
Clause 7.4: Purchasing
Clause 7.5.4: Customer Property
Clause 7.5.3: Identification and Traceability
Clauses 6.1: Provision of Resources, 6.3: Infrastructure, 6.4: Work Environment, 7.5.2: Validation of Processes for Production and Service Provision
Clause 8.2.4: Monitoring and Measurement of Product
Clause 7.6: Control of Monitoring and Measuring Devices
Clause 7.5.5: Preservation of Product
***Check* Cycle—Job Costing and Cash Flow Control**
Clause 8.2.2: Internal Audit
Clause 7.5.1: Control of Production and Service Provision
Clause 8.1: General Measurement, Analysis, and Improvement
Clause 8.2.3: Monitoring and Measurement of Processes
Clause 8.4: Analysis of Data
***Act* Cycle—Financial Feedback and Cost of Quality**
Clause 6.2: Human Resources
Clause 8.3: Control of Nonconforming Product
Clauses 8.5.2: Corrective Action with Customer Complaints, 8.5.3: Preventive Action

and general measurement, analysis and improvement; and control functions, such as control of documents and control of nonconforming product).

A similar diagram can be constructed for both the supplier (subcontractor) and the customer (interested parties). We can indicate the role of the organization (certified organization—you) when the value chain is extended to include this complete interorganizational flow. This unique functionality is demonstrated in Figure 5.7 (each arrow represents the set of executive functions, operational process, and control functions shown in Figure 5.6).

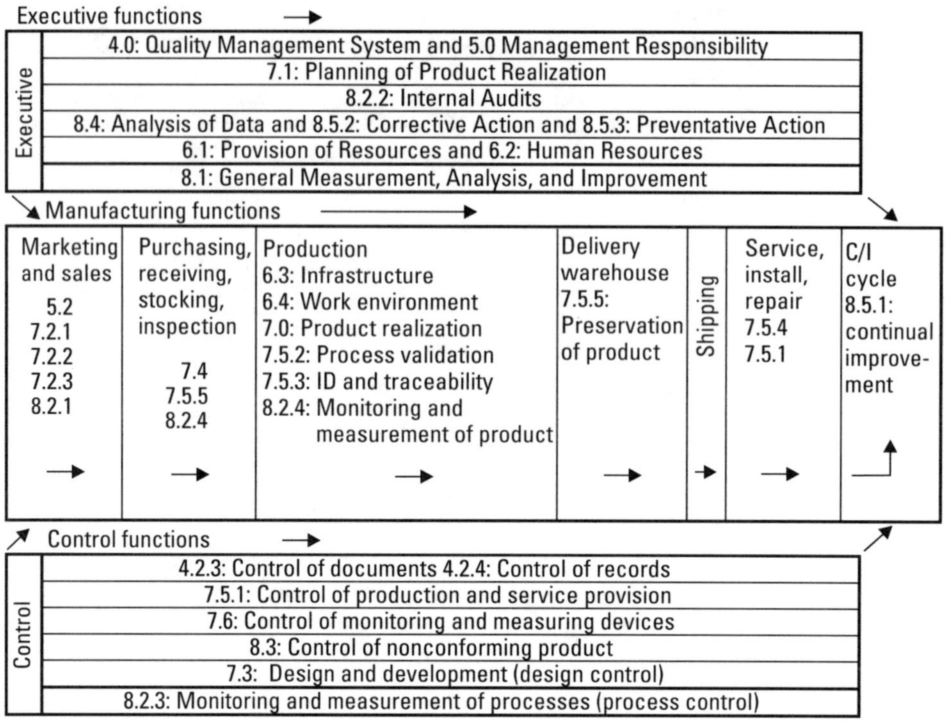

Figure 5.6 The ISO 9001:2000 QMS value chain.

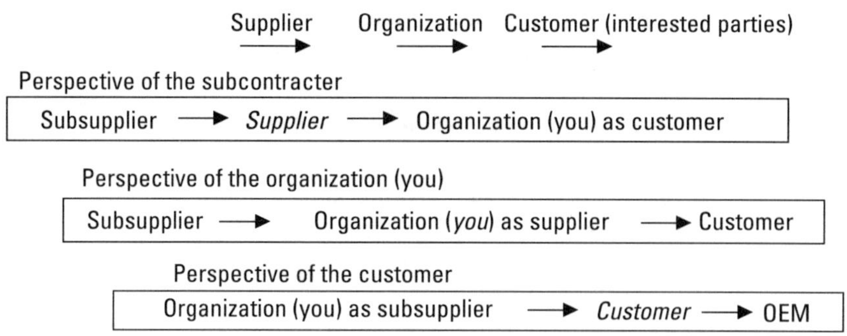

Figure 5.7 Value chain incorporating complete interorganizational flow.

As indicated above, the organization fulfills the role of customer, supplier, and subsupplier dependent upon where its various transactions occur in the chain. The middle diagram [perspective of the organization (you)] represents the basic terminology used in the Standard (i.e., the certified enterprise is termed the *organization,* those who provide resources to the organization are termed the *supplier,* and the customer is termed the *customer).*

The customer has been broadened to include interested parties that have "an interest in the performance or success of an organization" (refer to ISO 9000:2000, p. 10). Thus, the people in the organization can also be considered customers, similar to the TQM concept (i.e., anyone who receives a product).

A clarification of the terminology is offered in Table 5.14. The terminology equivalents are related to the Standard, its guidelines, and common industry usage.

5.12.5 According to Another Standard's Sequence

Unfortunately, the 1994 version was widely adopted as the baseline for other standards, such as QS-9000 and ISO 13485, and almost all of those standards will not be transformed into the 2000 format until 2003. As a result, the scenario—in which we decide to structure the manual against another standard's format—has already become a serious issue. We will deal with one such restructure because all of the other restructures are similar in form. The standard to be considered is the FDA/CGMP 820 National Standard (QSR) [32].

Table 5.14
Typical Supply-Chain Interface Terminology (* = Standard's Terminology)

*Supplier →	*Organization →	*Customer (interested party)
Subsupplier	Supplier	Customer
Subcontractor	Organization	Customer
Vendor	Contractor	Purchaser
Who supplies you	Certified organization (you)	Who you provide product/sell to
Business first party to you (you perform second-party audits on them)	Business first party to customer and business second party to supplier (you perform first-party audits on yourself)	Business second party to you (they perform second-party audits on you)
Metrology house	Producer	Consumer
Raw materials producer	Distributor	Client
Consultants	Importer	Independent supplier and producer
Contract supplier	Assembler	OEM house
Design house	Service organization	Sister division
Component manufacturer	Software house	Joint venture partner (may be internal or external to organization)
Private label manufacturer	Manufacturer	Organization's employees
	Design house (may be internal or external to organization)	Bankers
		Unions
		Society

5.12.5.1 Suggested Manual Format

We suggest that the manual be created as follows:

1. Template the manual in the form of the QSR (i.e., subparts and FDA/CGMP 820.*NNN* nomenclature).
2. Mathematician's rule: Like any good mathematician, first attempt to formulate the problem in terms of an already known solution—in this case, we can do a one-to-one relationship between the QSR and the ISO 9001:1994 version upon which the QSR is based, then relate the QSR to the Standard. Method of integration:
   - Table 5.7 already aligns the Standard's sections against the 1994 version.
   - Table 5.15 is an example of how one might integrate the results of Table 5.7 with the QSR.
   - Law of diminishing returns: Do not seek perfection—cross reference to the degree that makes sense. Follow the law of stupidity: "If it sounds stupid, it's stupid." Works every time.
3. Complete the manual with prescriptive, quality policy statements for every requirement of the QSR.
4. Evaluate each section of the completed QSR manual against the pertinent Standard requirements summarized in Table 5.15. Add in quality policy statements to the QSR sections wherever the Standard's requirements are not met.
5. Put Table 5.15, or an equivalent, into the manual as a cross-reference chart for reference or audit purposes.

5.12.5.2 Example of the Integrated QSR/Standard Quality Policy Statement

Let us apply Table 5.15 to a section of our manual that has been formatted according to the QSR. We have chosen Section 820.60: Identification as our sample of how the manual would sound with the Standard's requirements integrated into the QSR requirements. To create this paragraph we first responded to the QSR in detail and then include quality policy statements for the Standard's requirements (note the stress on process).

> *Section 820.60 Identification* (refer to the Standard's Par. 7.5.3 with regard to identification)
>
> The Excellent Company identifies product throughout the entire manufacturing cycle to maintain the integrity of its production process. The key

Table 5.15
Relationship Between the QSR, ISO 9001:1994, and ISO 9001:2000 National and International Standards

QSR Part 820 Sections	QSR 820 Titles	1994 Clauses	ISO 9001:1994 Titles	ISO 9001:2000 Clauses
820.5 plus 820.20 (d) and 820.20 (e)	Quality System	4.2	Quality System	4.1, 4.2.1, 4.2.2 5.4.2 (planning) 7.1 (planning)
820.20 minus 820.20 (d) and 820.20 (e)	Management; Responsibility	4.1	Management; Responsibility	5.1, 5.3, 5.4.1, 5.5.1, 5.5.2, 5.5.3(new) 5.6.1 6.1, 6.2.1, 8.5.1
820.22	Quality Audit	4.17	Internal Quality Audits	8.2.2, 8.2.3
820.25	Personnel	4.18	Training	6.2.2
820.30	Design Controls	4.4	Design Control	7.2.1, 7.3
820.40	Document Controls	4.5	Document and Data Control	4.2.3
820.50	Purchasing Controls	4.6	Purchasing	7.4.1, 7.4.2, 7.4.3
820.60 and 820.65	Identification; Traceability	4.8	Product Identification and Traceability	7.5.3
820.70 and 820.75	Production and Process Controls; Process Validation	4.9	Process Control	6.3, 6.4 7.5.1, 7.5.2
820.72	Inspection, Measuring, and Test Equipment	4.11	Control of Inspection, Measuring, and Test Equipment	7.6
820.80	Receiving, In-process, and Finished Device Acceptance	4.10	Inspection and Testing	7.1, 7.4.3, 7.5.3 8.1, 8.2.4
820.86	Acceptance Status	4.12	Inspection and Test Status	7.5.3
820.90	Nonconforming Product	4.13	Control of Nonconforming Product	8.3
820.100	Corrective and Preventive Action	4.14	Corrective and Preventive Action	8.5.2, 8.5.3
820.120, 820.130, 820.140, 820.150, 820.160, and 820.170	Device Labeling; Device Packaging; Handling; Storage; Distribution; Installation	4.15	Handling, Storage, Packaging, Preservation and Delivery	7.5.1, 7.5.5
820.180, 820.181, 820.184,	General Requirements: Device Master Record (DMR); Device History Record (DHR)	4.16	Control of Quality Records	4.2.4

Table 5.15 (continued)

QSR Part 820 Sections	QSR 820 Titles	1994 Clauses	ISO 9001:1994 Titles	ISO 9001:2000 Clauses
820.186, and 820.198	Quality System Record (QSR); Complaint Files	4.16	Control of Quality Records	4.2.4
820.200	Servicing	4.19	Servicing	7.5.1
820.250	Statistical Techniques	4.20	Statistical Techniques	8.1, 8.2.3, 8.2.4, 8.4
Not specifically addressed	Marketing and sales protocols are not addressed.	4.3	Contract Review	5.2 (customer focus) 7.2.1, 7.2.2, 7.2.3 8.2.1 (new)
Not specifically addressed	Usually processed in essentially the same system as noncustomer owned	4.7	Control of Customer Supplied Product	7.5.4

Source: [33]. Alert: Neither 5.5.3 or 8.2.1 are assigned in the Standard's cross-references.

method for hardware identification is by means of unique part numbers for each component, subassemblies, and final assemblies. In addition, serialization is provided at specified steps in the production process for all subassemblies and final assemblies. The part numbers and serial numbers are also used as identifiers during both return material and installation activities (refer to Doc. QA-001, entitled "Product Quality Plan").

Software, both embedded and hard-disk installed, is also identified by part number and serial number. Configuration management is the responsibility of the software engineering manager (refer to Doc Eng-002, entitled "Software Design Process").

Part numbers are assigned by engineering document control during creation of the bill of materials in the design and development phase (refer to Doc. Eng-001, entitled "Engineering Design Process").

Serialization of key assemblies is the responsibility of the manufacturing test department. All printed wiring assemblies are also serialized (refer to Doc. Mnf-001, entitled "Manufacturing Process").

5.12.5.3 The Special Case of a Manual:1994 Upgrade

As we discussed earlier, Table 5.8 is to be used when you add additional ISO 9001:2000 quality policy statements to a manual:1994. In other words, we are responding to the Standard's requirements using another Standard's sequence that in this case is the ISO 9001:1994 format.

Table 5.8 indicates the additional or expanded quality policy statements required to bring 1994 requirements in conformance with the Standard. So, for example, to bring Clause 4.20.1: Identification of Need of the 1994 Standard in conformance with the Standard, we would need to describe our policy with regard to the extent that statistical techniques are used in our QMS, and then ripple on down our policy into a tier II document. The result might look like the following in the manual:2000(20).

> *4.20.1 Statistical Techniques*: Statistical techniques are applied throughout the QMS processes. Such methods are described in document #ENG-2-005, entitled "Engineering Standards." The methods of data representation and analysis include the following:
>
> - Risk analysis, FMEA charts, and numerous computerized quantitative methods of analysis are used in engineering design.
> - Pareto and run charts are used in the analysis and actions taken with regard to quality objectives and corrective and preventive actions.
> - C=0 sampling techniques are used at incoming inspection, and run charts and histograms are used for in-process and final inspection.

5.12.6 Comparison of Sequences

The comparative analysis for the four sequences is diagrammed in Figure 5.8 and characterized in Table 5.16.

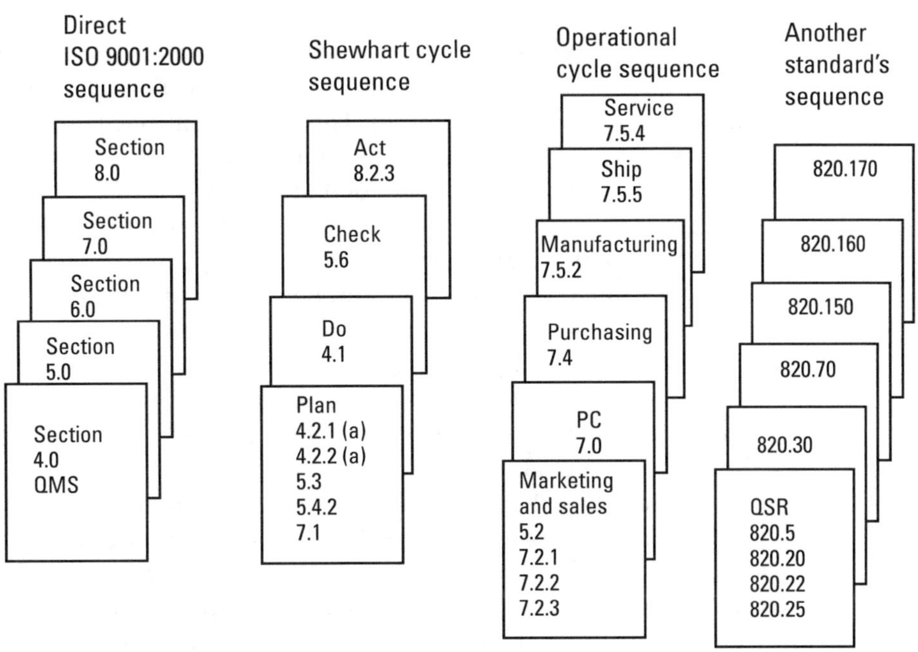

Figure 5.8 Diagram of quality manual potential sequences.

Table 5.16
Comparison of Quality Manual Content Attributes

Attributes	Direct Sequence	Shewhart Cycle	Another Standard's Sequence	Operational Cycle
Linkage to standard's clauses	Excellent	Good	Fair	Fair
Clarity of operational orientation	Excellent	Good	Good	Excellent
Continuous improvement cycle	Excellent	Excellent	Fair	Fair
Coverage of support functions	Excellent	Excellent	Excellent	Fair
Core competency clarity	Excellent	Good	Fair	Fair
Ease of auditing to Standard's clauses	Excellent	Good	Difficult	Fair
Overall	Excellent	Good	Fair	Fair

What we observe to date is that the direct sequence is the most common sequence chosen—although many previously certified suppliers have indicated that they would prefer to stay with the 1994 format discussed in Section 5.12.5.3. However, the Standard's format effectively serves to establish an operational division between structure (Section 4), management imperatives (Sections 5 and 6), the creation of product (Section 7), and the supplemental analytical functions (Section 8). This functionally elegant categorization is lost if one stays with the 1994 format (see earlier discussions of the Shewhart cycles for both the 1994 and 2000 Standards).

The Shewhart cycle is sort of a pocket-rocket approach to the direct sequence that already strongly reflects the plan-do-check-act cycle. It really doesn't offer more clarity and probably should not be considered a worthy candidate.

The operational cycle is strongly reflected in Section 7 of the Standard and, before you have accounted for all of the support and supplemental clauses, your final format will look a great deal like the Standard. So why not use a format that already exists?

Force fitting the Standard into another standard's format is readily doable with a little thought and a lot of effort. However, regardless of the other standard used, you will lose the beauty of the Standard's division of management functions into five clearly defined categories. In the case of previously created 1994 manuals, this may become the most popular approach, but it is retrogressive. In this regard, I have rated the use of the 1994 format as fair with respect to what I consider to be six prime attributes (see Table 5.16).

As for me, the direct sequence is number one on my list. It is somewhat redundant and certainly requires a great deal of interpretation. However, I feel it is the best we have for now and it provides a landscape that is unlimited in regard to core competency expression.

5.13 Manual Configurations

5.13.1 Two Unique Configurations

We will now demonstrate that regardless of which format is chosen for the manual's sections—direct sequence, Shewhart cycle, operational cycle, or use of another standard's sequence—there are effectively only two unique ways to design the manual's configuration:

1. Model I—stand-alone document that deals only with policy, scope, justified exclusions, the interaction between processes, and references to procedures;
2. Model II—integrated document that contains both policy, justified exclusions, the interaction between processes, and the procedures.

We maintain that any other form of documentation is a variation of one of these two basic forms. In addition, we will treat the manner in which linkage from the manual to lower-level documentation can be performed effectively.

The configuration decision is paramount in the choice of just where to place the quality policy statements. We have found it to be the primary source of conflict in manual design.

5.13.2 The Stand-Alone Configuration—Model I

In the case of a stand-alone manual—if the writer meticulously follows the criteria stated in the previous sections for the structure of quality policy statements—there will be no policy statements in the lower tier documents, because it will be unnecessary. Thus, we avoid redundancy, which is a nightmare in the maintenance of QMS documentation.

The stand-alone manual clearly references each lower tier document that it directly effects (e.g., the manual's Section 8.5.2: Corrective Action would send the reader to an SOP entitled "Corrective Action Procedure." This requirement for a referenced procedure is based upon both Par. 4.2.1(c)

Documentation Requirements—General of the Standard and Par. 4.2.2(b) Quality Manual of the standard.

5.13.2.1 Section References

This is normally done by inserting a reference document in each section of the Standard. Such an approach can produce a very complex and difficult navigational problem if there are many tier II documents. A typical example of this complex case would be in Section 7.5: Production and Service Provision, which might look like this: "The Excellent Corporation documents related to Production and Service Provision are to be found under the following:

1. Doc# 2-075-012, Production Control Procedures;
2. Doc# 2-075-011, Materials Control Procedures;
3. Doc# 2-075-004, Use of the Traveler Procedure;

...

12. Doc# 2-075-008, Release of Capital Equipment."

A way to avoid this difficulty is to create a process-related hub document that acts as a documentation flow center (refer to Table 11.1). When a hub document is used, the manual sends the reader to a handy universal bucket (hub) document (e.g., realization processes), and the hub document sends the reader to all of the documentation shown above.

5.13.2.2 Direct Sequence Stand-Alone Manuals

The exact form of the manual:1994 stand-alone manual has been almost entirely based on the direct ISO 9000 sequence, as described previously (i.e., a one-to-one correspondence to the numbering system of the Standard). This practice, to date, has held true for a manual:2000.

5.13.2.3 Quality Policy Statement Imperative

By definition, as defined in ISO 9000:2000, Par. 3.7.4, quality manual, we see that the manual is to specify the QMS in such as way as to direct and control the organization's quality imperatives. We also learn that the quality policy, as expressed by top management, is the way in which the organization receives the overall intentions and direction in relation to quality.

We conclude, therefore, that once the supplier has entitled a stand-alone document as the "Quality Manual"—regardless of the way a supplier chooses to label the contents of that manual—one thing we believe is clear is that all quality policy statements, to whatever level of detail is required, should be contained in the manual.

5.13.2.4 Primary Source of Inconsistency

The stand-alone manual configuration is the primary source of inconsistency in manual design because of the tendency to place policy statements not only in the manual but in lower tier documents as well.

5.13.2.5 Application to Third-Party Assessments

It is important to keep in mind that third-party assessors tend to go SHALL by SHALL through the manual on each clause, and each SHALL must be addressed. This exercise is prevalent at the document review, and it is far easier to accomplish this task when every SHALL is clearly addressed. If not, the result is a number of nonconformances.

Although many of the quality manual nonconformances at the document review are minors, they still require nonconformance reports (NCRs) and time to respond. The NCR could be a major finding if the process does not clearly exist.

In some cases (e.g., for a rewrite of 8.2: Monitoring and Measurement), the amount of rewrite could be so extensive that it would probably have to be done through the mail. This is certainly not the end of the world, but it could delay the certification by some number of weeks.

5.13.2.6 Corrective Actions for Dislocated Quality Policy Statements

If, as is often the case, the quality policy statements are located in lower tier documents, any one of the following corrective actions have been found to be acceptable:

- Cut and paste the statements into the quality manual (this minimizes redundancy);
- Copy the statements into the quality manual (this action does create redundancy but is not a nonconformance unless the lower tier document statements disagree);
- Include the lower tier document as part of the quality manual (this makes for a very heavy quality manual but is acceptable).

From an operational standpoint, although the placement of quality policy statements into lower tier documents instead of in the quality manual is commonplace, the fact that they are in a lower tier document that is not part of the quality manual is ineffective for the following reasons:

- Decision makers who require quality policy information would seldom (if ever) have the whole set of lower tier documents at their disposal;

- Lower tier documents generally contain proprietary information and are restricted in their distribution.
- The search for quality policy statements during a certification audit can lead to considerable waste of time and effort for all concerned.

I choose the stand-alone format. The desire to provide an adequate, consolidated statement of the organization's quality imperatives in one easily accessible and serviceable document is why essentially all suppliers have chosen the stand-alone form of the quality manual, and this is our recommendation. (Guidance on this subject is found in ISO 10013, "Guidelines for Developing Quality Manuals.")

5.13.3 The Integrated Manual Configuration—Model II

In an integrated manual, the quality policy statements and the lower tier documents—especially tier II—all appear in the same document. This is an approach that was commonly used some 30-plus years ago and can still be found in Mil-Q-9858A and FDA/CGMP-oriented manufacturing operations, as well as in the automotive industry.

In practice, in smaller companies, the manual and the set of SOPs are often distributed together in response to the Standard's Element 4.2.3(d): Control of Documents requirement to ensure that documentation is available where it is to be used. Let us consider the implications of the integrated manual approach [34].

Specifically, let us assume that we have chosen the integrated quality manual configuration. The manual will contain all of the quality policy statements required by the Standard in either of the following forms:

- Case I—process/SOP documents (tier II) that begin with a policy statement and are immediately followed by the process/procedural information (a joint document);
- Case II—abbreviated quality policy statements that reference an attachment that contains pertinent tier II as well as the rest of the required quality policy statements.

Figure 5.9 illustrates the form that such manuals could take.

As we noted earlier, such an approach is in full compliance with the Standard [i.e., Clause 4.2.2(b), which requires that the quality manual either includes procedures or refers to them].

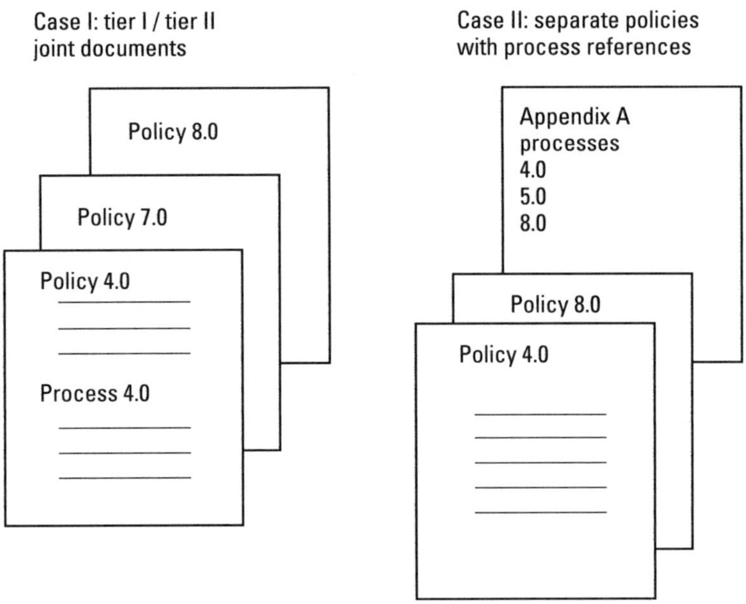

Figure 5.9 Schematic representation of possible integrated manuals.

5.13.3.1 Implementation

However, the implementation of such a decision is far from trivial, and we want to outline some of the more obvious issues:

1. When the assessors ask for the manual, they will expect to see the complete manual [i.e., either the full joint policy and tier II documented manual (case I) or all of the front-end text and the tier II appendix (case II). The appendix must include the full set of tier II documents that provide the rest of the required quality policy statements].
2. In the case of an appendix, they will expect to see the tier II attachment fully noted as part of the manual.
3. All attachments will need to be under document control; however, the list of tier II documents in the appendix need not show the revision dates because they will not expect you to keep more than one master list of documents. But they will need to know the full contents of any appendixes, either up front in the table of contents or as a part of the attachment.
4. When they check the distribution of controlled manuals, they will expect to see a complete document in the auditee's area. The "complete document" is the same as defined earlier (i.e., either the full jointly documented manual or the policy/appendix tier II manual).

5. When a specific quality policy statement is searched for—in response to a SHALL in the Standard—they can accept its presence anywhere in the integrated manual (i.e., front-end text or attachments).

6. They will expect to see a defining statement in Section 4: Quality Management System of the manual to the effect that the manual does consist of a set of either jointly documented tier I/tier II text or of abbreviated quality policy statements and associated tier II documents contained in an appendix.

Figure 5.10 offers a qualitative comparison of the stand-alone versus the integrated manual approach. The comparative analysis is indicated in an associated Table 5.17.

5.13.3.2 Discussion

As a result, the true difficulties with the integrated manual approach are with regard to the following:

- Large size;
- Diluted marketing orientation;
- Overkill in distribution.

It could get even more cumbersome in a multidivisional operation where it might happen that one manual was to be used in all divisions so that not only the corporate-level procedures but also the divisional-level procedures would be combined in one binder.

Figure 5.10 Graphic comparative analysis of configurations.

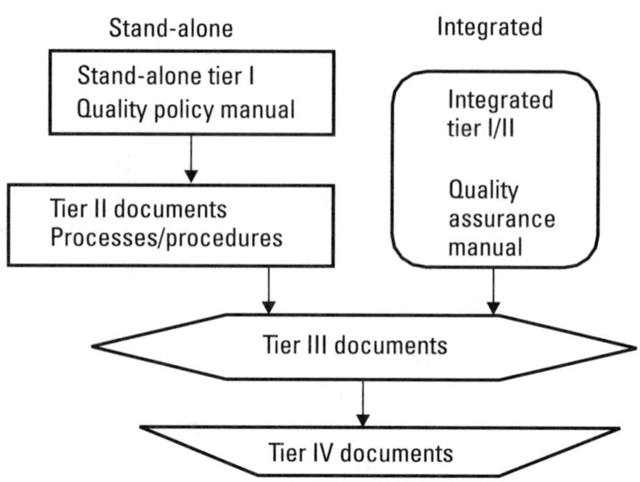

Table 5.17
Comparative Analysis of Configurations

Attributes	Stand-Alone Manual	Integrated Manual
Size scales with original size and complexity	Typically 30 to 75 pages without appendixes Only quality policy statements	Hub approach—200-plus pages without appendixes and forms. Can be considerably larger without a hub approach or for multisite manuals; only tier I/II documents for a single division or site
Redundancy	Usually has some with tier II and even tier III documents	Low probability
Linkage	By reference—somewhat difficult to learn	Direct linkage
As a teaching tool	Powerful with a big picture viewpoint	Somewhat limited because of narrower viewpoint
As a marketing tool	Outstanding—can clearly define organization's personality and image	Not really viable for marketing purposes—an issue of proprietary information and big-picture focus
Ease of auditing	Readily available information Quality policy statements readily found.	Readily available information Quality policy statements could be obscured
Ease of distribution	Can distribute manuals and tier II documents to custom fit the area	Each tier II area requires the entire manual to find its pertinent documentation
Overall	Most common in use	Not commonly found to date in ISO 9000 certifications

We have observed organizations' attempts to use an integrated manual approach by modifying their present Mil-Q-9858A or FDA/CGMP 820 documentations, but have eventually changed to the stand-alone model for ease of use and distribution. However, this approach has been used successfully in those cases where the supplier had little concern about the proprietary nature of the tier II documents and thus was not concerned about its distribution as part of a manual. We strongly recommend the use of a stand-alone quality manual for ease of use and flexibility.

5.14 Multidivisional Manuals

Our discussion can be readily expanded to include very large organizations with multidivisional requirements. Regardless of which basic configuration is chosen, the divisional documents would follow directly (e.g., the corporate manual would deal with policy at the highest level and the divisional manuals would respond to each corporate policy according to the specific operational characteristics of the division).

The corporate manual references the divisional manuals as appropriate. Some examples of the content of a divisional manual versus the corporate manual are summarized in Table 5.18.

A schematic of such a multidivisional manual structure, which follows the direct sequence pattern of the Standard, is shown in Figure 5.11.

5.14.1 ISO Management Review—Example of Labels

The block labeled "Corporate 5.0 quarterly reviews" represents the corporate response to the SHALLS in Section 5.0: Management Responsibility of the Standard. One of the key elements of this section has to do with the quarterly corporate management reviews and monthly divisional management reviews that are usually coordinated by the ISO 9000 management representative (see Clause 5.5.2: Management Representative). The assignment of the representative might be handled in the following manner at corporate and subsequently at division.

Table 5.18
Comparison of Corporate Versus Divisional Manuals

Subject	Corporate Manual	Divisional Manual
5.6: Management Review	Discusses the quarterly management review held at the corporate offices with divisional managers present	Discusses the monthly divisional management review that feeds the corporate quarterly review
	Describes the various corporate preparatory meetings that take place prior to the quarterly review	Also discusses the divisional preparatory meetings held locally prior to the monthly review
4.1: General QMS Requirements	Describes the entire QMS and references corporate standard operating procedures (CSOPs) or corporate process documents (CPDs)	Describes the response to the CSOPs via divisional work instructions (DWIs) Describes specific interface functions with corporate and other divisions
	Discusses the various ways that the divisions interface with both the corporate office and other divisions	Divisional SOPs or process documents are optional and are generally redundant
7.2: Customer-Related Processes	Describes the highest level marketing and sales policies and methods	Describes the method used at the divisional level to meet the corporate marketing and sales policies.
	References the corporate marketing and sales process manual	Describes the response to the corporate marketing and sales process by means of DWIs.

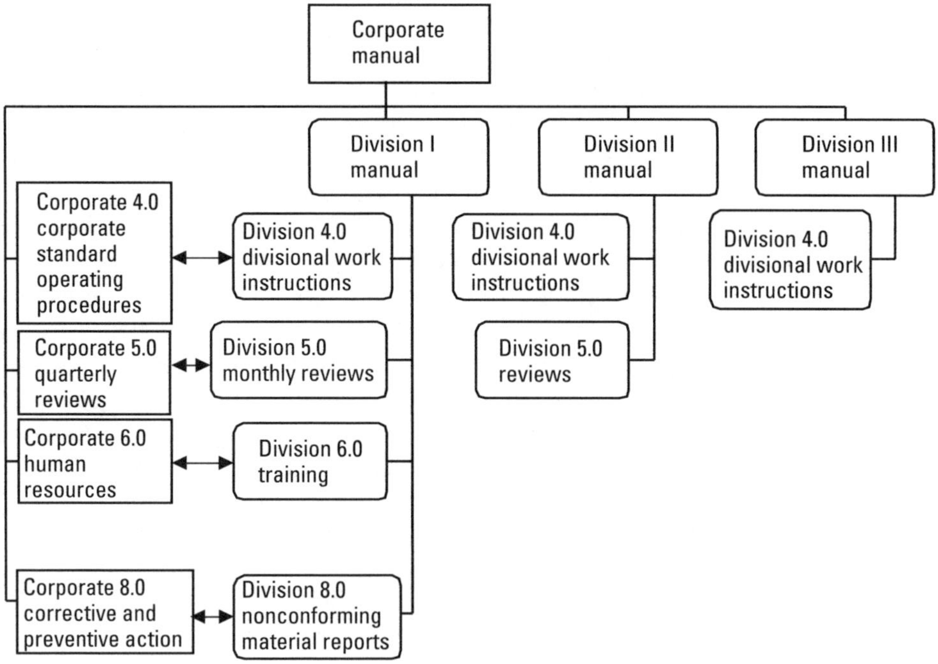

Figure 5.11 Possible configuration for a corporate/divisional integrated manual.

> Partial corporate assignment memo: "The president has designated the corporate director of quality assurance as the ISO 9000 management representative. A memo to this effect, which defines the additional duties and authority of this position, has been released by the president of the corporation and distributed to all employees of the corporation…"

The block labeled "Division 5.0 monthly reviews" represents the divisional response to the corporate assignment of an ISO 9000 management representative.

> Partial division assignment memo: "The general manager has designated the division's manager of quality assurance as the site ISO 9000 management representative who will coordinate such activities with the corporate director of quality assurance. Notice to all divisional employees was by means of the weekly quality improvement reviews held with the division's general manager and was made a part of the review minutes. The minutes included the additional duties and authority of this position..."

5.14.2 Summary and Conclusion

In this manner, each division responds in kind to the various corporate quality policy statements in order to form a cohesive and coherent body of corporate knowledge. Each division also shares the top-most level II documents (e.g., document control procedures, corrective and preventive action procedures, internal audit programs, and training manuals). Tier III and tier IV documents are designed expressly for use by a given division.

We have observed that the manual controversy occurs because although the supplier has chosen to write a stand-alone manual, the writers have been inconsistent with regard to the location of the quality policy statements that often appear in the lower tier documents.

We have found this tendency to confuse policy with process and process with procedure, independent of either organizational size or type of industry. In many cases, the SOPs are primarily statements of policy and the work instructions contain process descriptions.

We believe that a great deal of redundancy and misunderstanding could be avoided if authors always chose to use the following:

- Stand-alone manual configuration;
- Hub document approach;
- Stand-alone manual as the location of the quality policy statements.

5.15 Sector-Specific Manuals

5.15.1 The Accreditation Board Requirements

5.15.1.1 Purpose

Although it is unnecessary to be certified to ISO 9000 by an accredited registrar, accredited certifications are generally desirable due to their international recognition [35]. Unaccredited certification appears to be a choice for those who feel that ISO 9001:2000 is not viable for their organization but wish to continue their certification with a well-established registrar. A discussion of the accreditation process is covered in several texts, and any accredited registrar would welcome questions on this subject [36]. As a result, our purpose in this text is not to cover the details of accreditation but to instruct the reader in the ways accreditation board requirements affect the manual's structure.

5.15.1.2 Sector-Specific Requirements

Accreditation boards impact a supplier's certification process via sector-specific requirements that are passed on to the accredited registrar by means of memoranda backed up by a series of parallel standards. For example:

- EN45012: General Criteria for Certification Bodies Operating Quality System Certification;

- EN46001: Quality Systems—Medical Devices (particular requirements for the application of EN 29001; soon to be replaced by ISO 13485).

For example, requirements of this nature strongly affect certifications in the medical industry and to some degree in software design and manufacturing. In addition, the customer and/or the U.S. government can declare sector-specific conditions (e.g., as in the automotive industry's *QS-9000* requirements). In the case of QS-9000, GM, Ford, and Chrysler-Daimler operate as a team with the accreditation boards because the QS-9000 certification is impressed upon a certified ISO 9000 system.

5.15.1.3 Direct Applicability

What we wish to demonstrate is that all of the processes and methods discussed throughout our text are directly applicable to sector-specific assessments. In every case, the sequence and configuration techniques described previously will hold exactly in a sector-specific manual. In other words, the manner in which the Standard drives the lower level documents holds true regardless of the form of the Standard. The Standard could be ISO 9001, QS-9000, FDA/CGMP, or any other. It is still necessary to respond to each requirement.

5.15.2 Sector-Specific Quality Policy Statements

5.15.2.1 QS-9000 Example

The standard *Quality System Requirements QS-9000* was developed by the Chrysler/Ford/General Motors Supplier Quality Requirements Task Force to harmonize the several quality documents already in use by those companies. This standard is more than an interpretive guide in the assessment of automotive manufacturers. It must be adhered to in order for a company to receive a joint ISO 9001 or ISO 9002 and QS-9000 certification. The QS-9000 standard, which will require ISO 9001:2000 registration, is scheduled for revision to a new document in 2003 that carries the present identifier ISO/Draft TS 16949:2002 [37]. The technique demonstrated is invariant under this change.

The degree to which this automotive standard is to be applied is at the discretion of the registrar and their accreditation board based on interpretations provided by the International Automotive Sector Group (IASG) sanctioned by the Chrysler/Ford/General Motors. Such interpretations are published as special supplements in *Quality Systems Update,* a publication of the McGraw-Hill Companies, Fairfax, VA.

5.15.2.2 Specific QS-9000 SHALL

We will now consider a specific QS-9000 requirement and a typical response:

> Section II: Sector-Specific Requirements—Production Part Approval Process—General 1.1 states that suppliers are to fully comply with all requirements that are described in the production part approval process (PPAP) manual.

Notice that there is only one SHALL, but it is a big one. In fact, the rest of the requirement gives some general direction as to the protocols for subcontracted material, questions of part need, and approval. But it is necessary to go to the PPAP itself to determine the scope of the response. This is a 51-page document published by the Automotive Industry Action Group (AIAG), and we must assume that the supplier is extremely familiar with its content.

We can now write our quality policy statements in response to the PPAP directives. To do this, we first determine where the directives belong within the manual. This turns out to be easy, as it can go readily into an additional Section 9, entitled "Sector-Specific Requirements" (see boxed text).

Our response would then be found under Section 9 of the manual and would look something like the boxed text below, which should be an acceptable quality policy statement. In practice, a similar approach was fully accepted.

5.15.3 Current Good Manufacturing Practices Example

FDA/CGMP 21 CFR Part 820, Part VII, Quality System Regulation, Department of Health and Human Services, is a mixture of ISO 9001 and the previous Part 820. It is an integral part of any accredited assessment that involves medical devices. The degree to which this Standard is applied depends both on the class of devices manufactured and the discretion of the registrar and their accreditation board.

We will examine one of the Current Good Manufacturing Practices (CGMP) requirements and create a quality policy statement in response.

> Section 820.70 Production and process controls.
>
> (d) *Personnel*. Each manufacturer SHALL establish and maintain requirements for the health, cleanliness, personal practices, and clothing of personnel if contact between such personnel and product or environment could reasonably be expected to have an adverse effect on product quality.

When we analyze the requirement we see that there is only one SHALL but there are eight directives:

Excellent Corporation's Automotive Quality Policy Manual

Section 9: Sector-Specific Requirements

Production Part Approval Process

Procedure: Excellent's Standard practices and procedures used in the production part approval process (PPAP) is described in SOP# MNFG-2-21-001, entitled, "Production Part Approval Process." All procedures are based directly upon the AIAG publication entitled "Production Part Approval Process." Excellent contacts their customers directly when clarification is required with regard to this directive.

Responsibility: Excellent's quality assurance manager and quality control supervisor are responsible for the coordination and completion of the PPAP activity. This activity includes completion of the required documentation and submission of the appropriate PPAP documents to the customer.

Process: At Excellent, production part approval is always required prior to the first shipment of new parts, correction of discrepancies in shipped parts, and for modified parts managed and recorded within the ECO process. Notification of the customer by Excellent when parts are submitted for approval, unless waived by the customer, is the direct responsibility of the quality assurance manager. Excellent customers specify the submission level they require for each part on their initial purchase order. When no submission is specified, Excellent defaults to levels specified in the PPAP.

Submissions: Documents used in a PPAP submission include the part number, change level, drawing date, and identification as an Excellent part. Measurement system variation studies are conducted in accord with customer requirements. Special characteristics are referred to as critical, key, safety, significant, major, and minor. A Ppk index is used to determine acceptable levels of preliminary process capability.

Quality: Excellent's quality control department provides dimensional analysis, material tests, and performance test analysis based on the AIAG publication "Fundamental Statistical Process Control."

Records: The quality control department maintains all process records, engineering changes, and retains master samples."

1. Establish requirements for the health;
2. Establish requirements for the cleanliness;
3. Establish requirements for the personal practices;
4. Establish requirements for the clothing;
5. Maintain requirements for the health;
6. Maintain requirements for the cleanliness;
7. Maintain requirements for the personal practices;
8. Maintain requirements for the clothing.

We can now write our quality policy statements in response to the eight directives. To do this, we first determine where the directives belong within the manual. The Subpart G under which this requirement resides is entitled "Production and Process Controls," so it obviously belongs under 7.5.1 Control of Production and Service Provision of the Standard (see boxed text).

Our response would then be found under Section 7.5.1 of the manual and would look something like the boxed text below. Notice that we have responded in reasonable detail to all eight directives.

5.15.4 EN46001/ISO 13485 Example

The EN46001 Standard is published as a European standard and is entitled "Quality Systems—Medical Devices—Particular Requirements for the Application of EN ISO 9001." It is to be replaced in the near future by the ISO 13485 Standard published by the ISO Technical Committee 210. This new document is entitled "Quality Systems—Medical Devices—Particular Requirements for the Application of ISO 9001" [38]. The Standards are required as part of medical device certification for recognition by foreign parties (e.g., a requirement for CE marking or acceptance by Health Canada).

Both the EN46001and the ISO 13485 Standards require the same requirement for a medical device *file*. We will examine this particular requirement and create a quality policy statement in response.

> The 46001/13485 requirement under quality planning requires us to establish and maintain a *file* containing documents that define product specifications and quality system requirements—both process and quality assurance—for manufacturing, installation, and servicing activities, as they are appropriate—for each type or model of a medical device. Alternately, the reader can be referred to the location of the various documents.

Excellent Corporation's Medical Device Quality Policy Manual

Section 7.5.1: Production and Process Control

Personnel: The Excellent Corporation Standards for health, cleanliness, personal practices, and use of clothing apply to all personnel in contact with either medical components or finished goods in the clean room. SOP# Mnfg-2-07-011, entitled "Clean Room Dress Code and Regulations," defines the necessary procedures to ensure that these Standards are implemented and maintained overall by the manufacturing manager.

Dress Code: Company-issued uniforms are worn by all personnel while working with the components and products. For safety as well as sanitary reasons, strict rules apply with regard to shoes worn in manufacturing, assembly, packaging, warehouse, and laboratory areas. Also, strict rules apply with regard to hair and beard covers and the use of make-up and hand cream. The clean room supervisor is responsible for the strict adherence by all clean room employees to this Standard.

Facilities: Locker, wash, and coat rooms—all employees and visitors to the manufacturing and assembly areas use approved entrances and exits. Lockers are available so that street clothes and personal items can be stored before entering the manufacturing areas. Personal cleanliness is required after rest room use. Area supervisors monitor and enforce all aspects of this Standard.

When we analyze this requirement we see that there is only one SHALL, but there are three directives with six subdirectives—or nine total directives. That is, the file is to contain (or reference) documents that define the following for each medical device with regard to manufacturing, installation, and servicing, as appropriate:

1. Product specifications to manufacture;
2. Product specifications to install;
3. Product specifications to service;
4. Quality system process requirements to manufacture;

5. Quality system process requirements to install;
6. Quality system process requirements to service;
7. Quality-assurance requirements to manufacture;
8. Quality-assurance requirements to install;
9. Quality-assurance requirements to service.

We can now write our quality policy statements in response to the nine directives. To do this, we first determine where the directives belong within the manual.

The technical file (as it is usually referred to) can be described in either section 4 as a record or in Section 7 as part of the design phase. Because it is actually a collection of documents, we choose to include it in Clause 4.2.4: Control of Records.

Our response would then be found under Clause 4.2.4 of the manual and would look something like the boxed text below. Notice that we have responded in reasonable detail to all nine directives (i.e., we cover manufacturing, installation, and service as a function of the design, manufacturing, and quality-assurance processes).

5.16 Potential Manual Readership

We now enter into the realm in which our models and design theories come together to form documentation that relate in some incomplete way to reality. We must deal with an entity—the enterprise—that performs daily based on the interweaving of both effective and affective employee behavior in juxtaposition with an external environment that is essentially unpredictable—the marketplace. Effective behavior results from employee agreement with policy, process, procedure, and formats; affective behavior results from the psychological orientation of the employee due to both enterprise and general life stresses. The degree to which the employee accepts the effective portion of this duality is greatly impacted by the employee's affective orientation.

Unlike an electronic circuit board, we cannot hook up probes on our employees and measure currents and voltages that indicate our success or failure to meet specification. Our problem is that we are not only limited in our ability to fully project the dynamic behavior of our employees as they use the documents, but we cannot even predict the impact of our written words on our customers. The manual is a prime example of this weakness. However, as we analyze the problem, we can offer some effective solutions.

Excellent Corporation's Quality Manual

Part 4.2.4.1: Technical File Compilation and Records

Procedure: Excellent's procedure to create a technical file for each product that requires medical device documentation is described in document MEDDEV-2-04-001, entitled "Medical Device Reporting Procedures."

The technical file is presented in the form of a table that indicates the specific requirement, the name of the document designed to meet this requirement, the location of the document, and the document's owner. The format of the table is as follows:

Technical File for Product: ____________________

Content Requirement(s)	Documentation Responsibility	Document Location	Document Owner
1. Complete product specifications	Design engineering group	Exact file area	Specify individual
2. Installation process manual	Customer support group	Exact file area	Specify individual
3. Servicing process manual	Customer support group	Exact file area	Specify individual
4. Description of device with variants	Continuous engineering group	Online files	Specify engineering, IT, and document control team
5. Manufacturing process document	Operations	Online files	Specify IT and manufacturing team
6. Quality assurance process manual	Quality assurance and regulatory affairs (QA&RA)	Exact file areas and online files	Specify operations and quality assurance team
7. Master list of Standards	QA&RA	Exact file area	Specify one individual
8. Master list of all documents with QMS	Document control	Exact file areas and online files	Specify individual
9. Declaration of conformity	QA&RA	Exact file area	Specify individual
10. Purpose and objective of file	QA&RA	Exact file area	Specify individual
11. Intended use, classification, and rules	QA&RA	Exact file area	Specify individual
12. Essential requirements	QA&RA and design and development (D&D) engineering	Exact file area	Specify team
13. Risk analysis	QA&RA and D&D engineering	Exact file area	Specify team
14. Clinical data	QA&RA and D&D engineering	Exact file area	Specify team
15. Justify choice of materials with packaging	QA&RA and D&D engineering	Exact file area	Specify team

Excellent Corporation's Quality Manual Technical File for Product (continued)

Content Requirement(s)	Documentation Responsibility	Document Location	Document Owner
16. Labeling samples	QA&RA and D&D engineering	Exact file area	Specify team
17. Clinical data	QA&RA and D&D engineering	Exact file area	Specify team
18. Sterility methods	QA&RA and D&D engineering	Exact file area	Specify team
19. Repeatability Methods	QA&RA and D&D engineering	Exact file area	Specify team
20. Method to notify the notified body on changes	QA&RA	Exact file area	Specify individual
21. Medical device reporting processes	QA&RA	Exact file area	Specify individual
22. Pertinent test data	QA&RA and D&D engineering	Exact file area	Specify team
23. Marketing literature	QA&RA and marketing and sales	Exact file area	Specify team
24. Authorized European representative	QA&RA	Exact file area	Specify individual

Specifically, the manual is the most difficult ISO 9000 system document to write. Of all the ISO 9000 documents, it must appeal to the widest set of readers. As a result, the purpose of the manual must be carefully couched in terms of its users. We can classify the potential readers of the manual as follows (see Table 5.19).

As can be seen from Table 5.19, the potential readership for the manual is extremely diverse and must comply with an ever-expanding set of user needs.

5.17 Manual Objectives

However, from the perspective of readership, we can at least attempt to define the overall objectives of the manual:

- To clearly describe the organization's QMS with enough detail to make it useful for a very wide range of readers;
- To respond to each requirement of the Standard so that the defined system has the potential to achieve the full benefits of continuous/continual improvement—intrinsic within the Standard;

Table 5.19
Classification of Potential Manual Readers

Potential Readers	Includes	Reader Decision Needs
Customers/clients/ partners	Executives	To audit or not to audit
	Purchasing agents	To buy or not to buy
	Quality assurance managers	To invest or not to invest
	Operations managers	Initially based on the scope and completeness of the quality manual
	Distributors	
	Sales representatives	
	Investors	
	Interdivisional organizations	
Employees	Executives	Is the organization really committed to quality?
	Managers at all levels	How can I participate?
	Engineers	What is expected of me as a quality person?
	Supervisors	What are the quality rules of the house?
	Technicians	
	Assemblers	
	Buyers	
	Marketing and sales personnel	
	Internal quality auditors	
Subsuppliers	Subcontractors	What level of quality is required?
	Vendors	How will I be measured?
	Interdivisional organizations	What type of supplier audit can I expect?
		Will I be rewarded for my work?
Third-party ISO 9000 evaluators	Assessors	Degree of compliance to SHALLS
	Registrars	Dedication of top management
	Accreditation boards	Potential for continual improvement and effectiveness of the quality system
	Third-party experts	The extent to which quantitative methods are used to measure effectiveness

- To write the manual from the customer/client's perspective as the primary issue for clarity and preciseness.

ISO 10013:1995, Par. 4.2.1, clearly describes additional objectives of the manual (e.g., its use in audits, training, and implementation). What we define as objectives is from a more holistic business perspective and one that ingrains the organization's personality.

New Customers Of all the readers, it is the customer, and in particular the new customer, who must rely on the manual as a primary source of organizational quality competence. Every other reader has some degree of organizational knowledge either through a series of business transactions or through audits.

The new customer looks at the organization with fresh eyes based on a particular technological and financial viewpoint. As a result, if the manual is written at a level of clarity sufficient to satisfy the new customer, it will certainly satisfy any other reader.

This assumption implies that there is sufficient detail to satisfy the needs of a still-diverse set of readers (e.g., executives, purchasing agents, quality assurance-managers, investors, and interdivisional organizations).

Endnotes

[1] Par. 3.7.4 of ISO 9000:2000.

[2] Par. 4.2.2 of ISO 9001:2000.

[3] ISO 9001:2000, Clause 4.2.2.

[4] "ISO Guidance on the Terminology Used in ISO 9001:2000 and ISO 9004:2000," *Quality Systems Update,* Vol. 11, No. 2, February 2001.

[5] See, for example, Kannan, V., et al., "Tools and Techniques of Quality Management: An Empirical Investigation of Their Impact on Performance," *QMJ,* Vol. 6, No. 3, 1999.

[6] See, for example, Russell, J. P., *The Quality Audit Handbook, Second Edition,* Milwaukee, WI: ASQ Quality Press, 2000.

[7] See, for example, "First Interpretations Stall Over Possible Lapse in Process," *Quality Systems Update,* Volume 11, Number 11, November 2001, at http://www.qsuonline.com.

[8] This observation is supported in part by data presented on p. 20 of *Quality,* (January 1997), which notes that, in general, larger companies have been registered longer than smaller companies.

[9] The *Registered Company Directory, North America, 1999,* McGraw-Hill Companies, publisher of *Quality Systems Update,* Columbus, OH (e-mail: customer.service@mcgraw-hill.com) lists the profiles of 84 accredited registrars and four unaccredited registrars in North America.

[10] See, for example, Harwood, David, "Integrating Quality, Health, Safety, and Environmental (QHS&E) Management Systems with Central Document Control," *Proc. Eighth Annual International Conference on ISO 9000,* February 7–8, 2000, Orlando, Florida.

[11] Although there is a plethora of books on these subjects, Chapter 25 of Kotler, Philip, *Marketing Management, Sixth Edition*, Englewood Cliffs, NJ: Prentice Hall, 1988, provides a excellent and surprisingly up-to-date summary of the role of finance and marketing in the corrective and preventive action process.

[12] See, for example, Campanella, Jack (ed.), *Principles of Quality Costs, Second Edition*, Milwaukee, WI: ASQ Quality Press, 1990.

[13] See for example: Crist, Roger, "E-Documenting for Better Control," *Quality Digest*, March 2001, p. 41 (http://www.qualitydigest.com). Also, monthly publication *Managing Automation*, Thomas Publishing Co., New York (http://www.managingautomation.com), is a cornucopia of information related to the diverse methods in use today to automate enterprises (e.g., Neil, Stephanie, "The Untethered Terrain of Wireless," *Managing Automation*, April 2001, p. 26, and Rice, Valerie, "Success with Web Customer Service Depends on More than Technology," p. 34. The June 2001 issue of *Managing Automation* lists 96 companies that sell manufacturing software.

[14] There is a difference of opinion on whether the quality manual requires quality policy statements for each SHALL of the Standard. This remains a fundamental issue in manual structure. Our interpretation is based upon Annex C of the ISO 10013:1995 Guidelines for Developing Quality Manuals. In this annex, the example given is the set of quality policy statements that form a response to Element 4.17: Internal Quality Audits of the 1994 Standard. The ISO "Guidance on the Terminology Used in ISO 9001:2000 and ISO 9004:2000" defines "Shall (Shall not): Used to indicate a requirement strictly to be followed in order to conform to the standard and from that no deviation is permitted. Do not use 'may' or 'may not' as alternatives." (http://www.iso.ch).

[15] Edward P. Link publishes a pocket-sized booklet that stresses the value of reduced variation as it relates to ISO 9001:2000 [i.e., *An Audit of the System, Not of the People*, Quality Pursuit, Inc., Rochester, NY, tel: (716) 586–2680].

[16] "ISO Guidance on the Terminology Used in ISO 9001:2000 and ISO 9004:2000," *Quality Systems Update*, Vol. 11, No. 2, February 2001.

[17] Refer to "QS-9000 Quality System Requirements," obtainable from AIAG, (810) 358-3003 for a sector specific definition and usage of "shall" and "should."

[18] *Webster's Encyclopedic Unabridged Dictionary of the English Language*, New York: Gramercy Books, 1996.

[19] Par. 5.6 Management Review.

[20] The controversy on quality manual design appears to have reached a level of concern sufficient to cause publications on the subject (e.g., Balano, Richard, "How to Avoid Creating the Dreaded 'Big Honkin' Binder," *Quality Progress*, March 1997, p. 152; and some years later, Wright, Tony, "ISO 9001 Without Tears," *Quality Progress*, August 2001, p. 57.

[21] See, for example, Wright, Tony, "ISO 9001 Without Tears," *Quality Progress*, August 2001, p. 57.

[22] See Shipley, David, "Destination: ISO 9001," *Quality Progress,* March 2002, p. 32, for an example of how one company faced this upgrade issue.

[23] ANS/ISO/ASQ Q9001-2000 Standard, pp. 19 and 20—refer to those pages for all 1994 clauses.

[24] "Top Three British Registrars Forecast Registration Push," *Quality Systems Update,* Vol. 11, No. 11, November 2001, and "Super Committee Tackles ISO 9000 Issues, Looks for Answers," *Quality Systems Update,* Vol. 12, No. 5, May 2002, at http://www.qsuonline.com.

[25] This assumption may eventually change because many suppliers state that they would prefer to stay with the manual:1994 format. It is subject of intense discussion at this writing. My experience to date with 12 manuals indicates that the 2000 format was preferred, so once people start working the problem they may change their preference. Of 12, 3 stayed with 20 clauses.

[26] For more information on the way a registrar views a client, see, for example, Taylor, C. Michael, "Meet the Registrar: Firsthand Accounts of ISO 9000 Success from the Registration Source," ASQ Publications, 1997, item: H0967, ISBN 0-87389-423-3.

[27] The 90% rule is not meant to negate the desire for perfection. It does emphasize the reality of life among those who already have a full-time job in addition to their ISO functions. At any time prior to certification, during certification, and after certification, there will always be about 10% to 20% of the system under revision. That is called *continuous improvement.* Any attempt to beat this fact of life will be met with extreme frustration followed by rebellion. Minor nonconformities due to fine tuning are a constant of the motion in ISO.

[28] The publication of the ANSI/ISO 14001-1996, Environmental Management Systems—Specification With Guidance for Use document is based upon the Shewhart cycle. ISO 9001:2000 then followed with its reconfiguration to essentially the Shewhart cycle.

[29] Based on the work of Dr. Anthony F. Costonis, president and founder of Corporate Development Services, Inc., of Lynnfield, MA, http://www.corpdevelopment.com. Although the Shewhart cycle was used to certify a client a number of years before the 2000 version release, the assessors had little trouble with the QMS structure because of a detailed cross-reference chart.

[30] See, for example, the *ASQ Quality Press Publications Catalog,* Fall/Winter 2001, pp. 92–95, at http://www.asq.org.

[31] The concept of a value chain was first introduced by Michael E. Porter in a series of business text books (e.g., *Competitive Advantage,* New York: The Free Press, 1985). The idea of "chains" finds many outlets (e.g., Troczynski, Tom, "The Quality Chain," *Quality Progress,* September 1996, p. 208.

[32] Quality System Regulation, Part VII, Department of Health and Human Services, Food and Drug Administration, 21 CFR, Part 820, Medical Devices; Current Good Manufacturing Practice; Final Rule, Monday, October 7, 1996.

[33] ANS/ISO/ASQ Q9001-2000 Standard, pp. 19 and 20—refer to those pages for all 1994 clauses and FDA/CGMP 820 National Standard (QSR).

[34] We do not wish to imply a dislike of the integrated manual. This is not the case because in my own practice I have designed integrated manual systems when it was required. However, as noted, this form of manual requires considerable thought with regard to ease of use, distribution, and maintenance.

[35] ISO 9000 astute purchasing agents, who are already on guard because it is not unusual for an ISO 9000–certified company to ship nonconforming product (the complexity of real-life shipping schedules and rapid technical changes are always with us), may tend to value less a nonaccredited registration. However, in most cases, purchasing agents tend to do what is best for their companies and get what they need when they need it. There are no simple answers in this case.

[36] See, for example, Tiratto, Joseph, "Registrar Accreditation," *The ISO 9000 Handbook, Second Edition*, Fairfax, VA: McGraw-Hill Companies. Another McGraw-Hill publication, *ISO 9000 Registered Company Directory, North America*, provides a running list of accredited and unaccredited registrars on a quarterly basis.

[37] Munro, Dr. Roderick A., "Future of APQP and PPAP in Doubt," *Quality*, January 2002, p. 28, at http://www.qualitymag.com, "Next TS 16495 to Require ISO 9001:2000 Registration," *Quality Systems Update*, October 2001, p. 1, and "The New Global Automotive Quality Management Standard ISO/TS 16949," May 2002, p. 17, at http://www.qsuonline.com.

[38] *ISO 9000 and 13485 Essentials: A Practical Handbook for Implementing the ISO 9000 and 13485 Standards for Medical Device Manufacturers, Second Edition*, CSA, Ontario, Canada, tel: (800) 463–6727.

CHAPTER

6

Contents

Process Document Design

6.1 The Process Document

The tier II document is expressed in many different ways, all of which are identical. For example:

- SOP;
- Process document;
- Hub document;
- Quality-assurance procedure;
- Quality or control plan.

The role that the process document plays in the QMS is to describe the time-dependent behavior of the system after it has been defined in terms of quality policy statements. The importance of the process document cannot be overly emphasized. It is actually the first document that should be drafted prior to any other, including the quality manual (or alternately, the quality policy manual or quality system manual).

6.1.1 The Critical Development of Processes

Because the identification of processes, their sequencing and interaction, and a description of such interactions is the most dramatic revision to the 1994 version, it requires that we carefully analyze the way in that this critical requirement can be responded to effectively.

Let us consider a typical manufacturing company and what its total marketing and sales to after-sales service process could look like as our identification of processes [refer to Par. 4.1(a) of the Standard]. This concept is shown in Table 6.1. The process is described in tabular form although we could have used a flow chart graphic.

We have chosen to show the business process discussion as part of the manual because a description of the interaction between the processes is required to be within the manual. The illustration includes the formatting technique used to create this integrated text and graphics approach. The manual has been formatted to exactly match the Standard's nomenclature so the text begins with 4.1: Requirements—Apogee has chosen to define its operational processes in terms of … . As demonstrated, the integrated example covers the requirements found in Par. 4.1(a), 4.1(b), and 4.2.2(c) of the Standard.

Business/quality objectives are used to measure progress at the corporate and departmental levels. The sequence and interaction of the processes are identified in terms of the input and output activities in that the output of one phase becomes the input to the next phase. In this manner, the manual describes the interaction between the processes of the QMS as required by Par. 4.2.2(c) of the Standard.

The process document is the "homework" part of the QMS design. It is the research mechanism that establishes whether we really understand the organization's dynamics. It is the document, as incomplete as it might be, that models the organization's ability to manage change and continuous/continual improvement. There is no specific way to write the document, and we will demonstrate two methods that can be used to compose the initial draft (i.e., a process table approach and a cyclic flow chart approach.

In the first exercise, the Steward for Element 7.5.1: Control of Production and Service Provision has brought her team together and created a department-to-department process document that carefully maps the interfaces or "hand-offs" from one department to another. The results are shown in Table 6.2.

In the second exercise, the Steward for Element 5.0: Management Responsibility has brought the executive group together and created a cyclic flow chart to diagram the entire development process. The results are shown in Figure 6.1.

6.1.2 Process Document Application

Another application of the recommended process document in QMS structures is illustrated in Figure 6.2. In this figure, each circle directly attached to

Table 6.1
Apogee E&M, Inc., Business Process Chart*

4.1 Requirements Apogee has chosen to define its operational processes in terms of core competencies that represent the various phases of the business cycle. The business process is summarized in the following table.

Core Phase	Core Activity	Managed by...	Inputs	Outputs	Resources/ Controls	Links to Document
1.0	Board of directors	Chairman of the board	Investors Quarterly operations reports	Annual report Approved Budgets and Corporate Business Plan	Stock Approved budgets	Annual Apogee budget and business plan
2.0	Executive office	President and chief executive officer Chief financial officer Chief information officer Human resources QA&RA	Annual Apogee corporate budget and business plan Department monthly reports	Corporate and departmental objectives and goals and budgets	Stock plan Budgets Objectives Management review	Departmental budgets and business plans (with objectives and metrics) Employee handbook
3.0	Marketing and sales	Vice president (VP) of marketing and sales	Marketing and sales budgets and business plans Customer complaint status	Sales forecast Profit and loss (P&L) Distributors Representatives New product requirements Return material authorizations (RMAs)	Sales force Budgets/ objectives Advertising Brochures Price lists	Marketing and sales process document Product brochures Customer service process document
4.0	Research, development, technology, and engineering (RDT&E)	VP of RDT&E	RDT&E budgets and business plans New product requirements	Prototypes Product specification Regulatory compliance Test equipment Documentation	Engineering staff Budgets/ objectives Capital equipment Project plans	RDT&E process document Engineering standards manual

Table 6.1 (continued)

Core Phase	Core Activity	Managed by...	Inputs	Outputs	Resources/ Controls	Links to Document
5.0	Operations	VP of operations	Operations budgets and business plans Production transfer package	Manufactured products Private labels Warehousing Facilities management Outsource management RMA repairs	Operations staff Budgets/ objectives Yield and scrap analysis Preventive maintenance	Operations process document Device history records
6.0	Customer service	Customer-service manager	Customer-service budgets and business plans RMAs Product specifications	Spare parts and repairs P&L Repairs Installations Customer-complaint management	Customer-service staff Budgets/ objectives Field service personnel Repair facilities	Customer-service process document Customer-service brochures and price lists

* The business process chart defines responsibility for each core competency of the operation and references the appropriate next tier documentation required to implement an effective quality management system.

the executive process represents a process document, including the executive process itself.

Figure 6.3 indicates how the executive process document sends the reader to specific tier III procedural documents. The same type of structure can be generated for each of the process documents illustrated in Figure 6.2.

For example, the marketing and sales process document would send the reader to such procedures as marketing guidelines, program management guidelines, OEM sales guidelines, distributor agreements, and order entry.

The engineering design process to hardware design guidelines, software design guidelines, engineering document control, engineering standards, and design transfers to manufacturing. The manufacturing process document to inspection and testing, preventive maintenance, facilities, purchasing, control plans, control of nonconforming material and product, ESD control, and packaging. The customer service process document to installation guidelines, returned goods, and field service contract agreements.

In this manner, there is a clearly established link between the quality manual prescriptive quality policy statements, the organization's core competencies expressed as process documents, and the procedural documentation.

Table 6.2
Excellent's Operational Processes

Stages	Description	Activities and Hub Documentation
1	Marketing and sales (M&S) procurement process	The vice president of M&S manages the procurement of contracts via the quoting process and manages the use of advertising and brochures to increase market share. S&M 001.
2	Design and development process	The vice president of engineering manages the design and development of new and modified products as required through the procurement process. Manufacturing release packages are prepared by the engineering department. Eng 001.
3	M&S purchase order review process	The vice president of M&S supports the manufacturing process via the management of customer purchase orders. Sales and marketing supplies the release package to production control. S&M 002.
4	The build plan	The vice president of manufacturing prepares the build schedules and review with purchasing. MNF 001, Stage One.
5	Purchasing	The purchasing manager negotiates and procures material required to meet kit schedules. PUR 001, Stage One.
6	Receipt of raw material	The receiving and shipping department verify and receive raw material, and quality assurance runs acceptance testing. MNF 001, Stage 2; and QA 002.
7	Stocking raw material	Production control stocks, releases, and inventories raw material. MNF 001, Stage 3.
8	Kitting	The materials department kits raw material and transfers the kits to production control. MNF 001, Stage 4.
9	Assembly and test	Production assembles and tests boards and integrates box level product. MNF 001, Stage 4; and WI series 1-001 through 1-087.
10	Shipment to customer	The shipping department verifies customer order/pick ticket, final configuration and testing, and packaging and shipping documentation, and delivers product to customer. MNF 001, Stage 5; and W/I series 2-001 through 2-016.
11	Customer service	The customer-service manager manages the return of products, their repair, and the analysis of nonconformities in conjunction with the quality-assurance manager.

6.2 The Trouble with Tier II

Explicitly, a process document is not a mandatory document. Implicitly, it can be—if it is in the form of a procedure that is required. For example, the audit process could be documented in the form of a SOP that would then be

Figure 6.1 Excellent's total business processes.

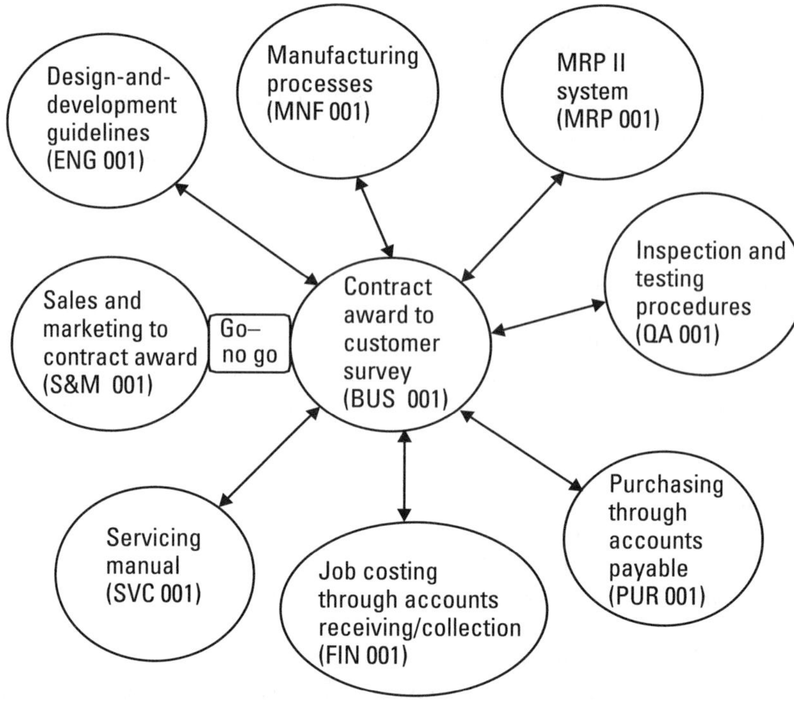

Figure 6.2 Wolf Telecommunications, Inc.'s (WTI) business process.

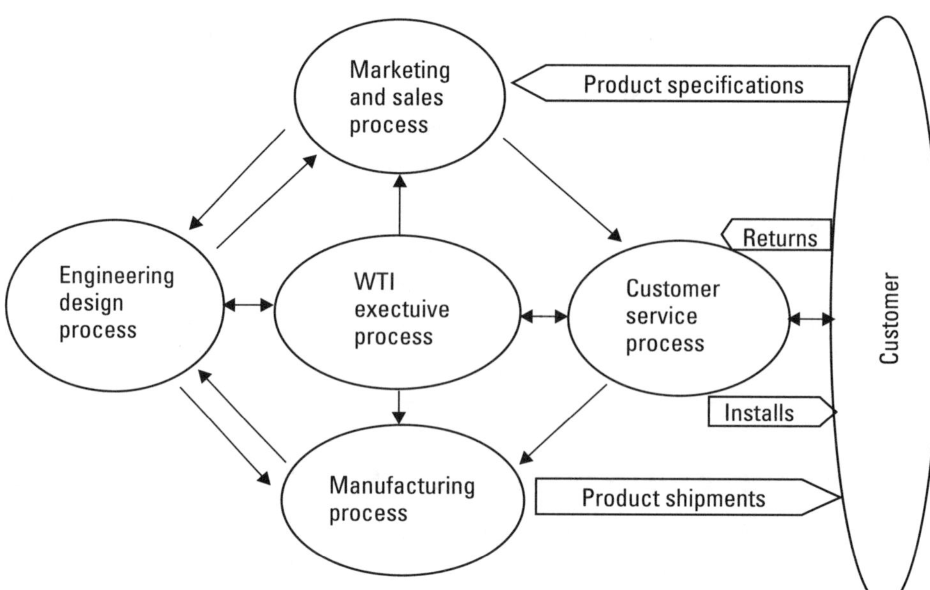

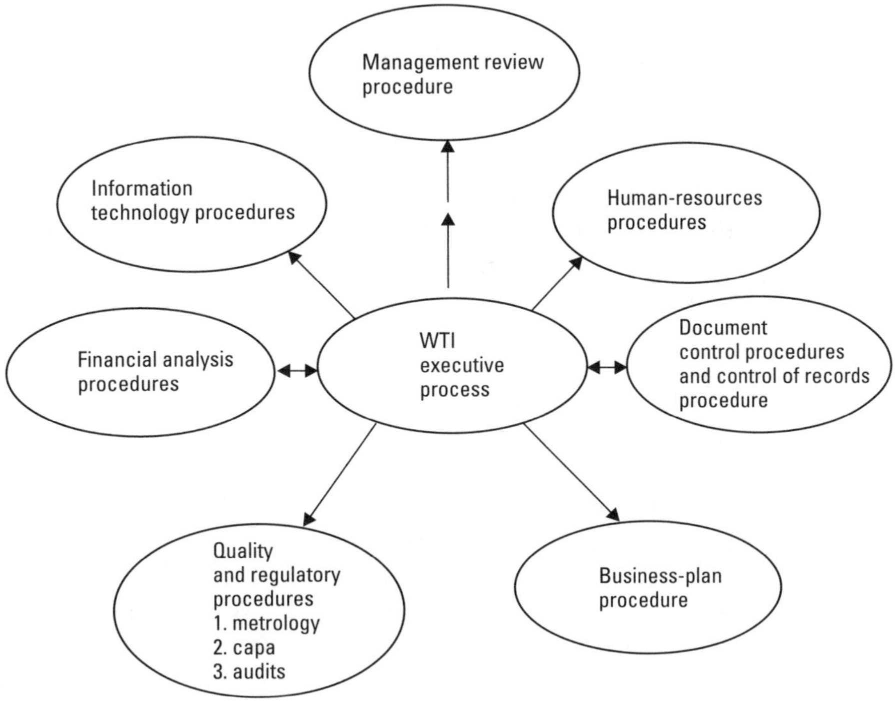

Figure 6.3 Wolf Telecommunications, Inc.'s (WTI) executive process with associated procedures.

mandatory because an audit procedure is mandatory. This vagueness is not new to the 2000 version; it has always been there and has always been confusing to all QMS developers.

However, several requirements and definitions help to demystify the form of tier II documentation. Such inputs can serve to include the concept of a process document more clearly into the ISO terminology.

First, we must examine the definition of a procedure. With reference to ISO 9000:2000, Par. 3.4.5, a procedure is a document that tells you how to accomplish either an activity or a process. In other words, if you want to create a process document, you can call the document a procedure. The common terminology ranges from standard operating procedure, to quality systems procedure, to quality-assurance procedure. The document will then fit into the ISO terminology.

There is another bug in the ointment that is a throwback to the 1987 first release. Procedures can be documented or not. What we did in those days to resolve this issue was to interview several people running the same procedure to indicate either that it was being done differently by different people or it was not. The advent of a multitude of procedures and work instructions is indicative of what was discovered.

However, this requirement, which appears as a note under Par. 3.4.5 in the vocabulary, is a real issue that must be considered carefully. For example, many a machine shop has extremely well-qualified and experienced machinists who perform a multitude of complex tasks without written procedures. To require written procedures in this case would be a waste of resources. As long as both the inputs and outputs of the machining process are controlled and the appropriate records are kept, there is no sensible reason to document how the machinists set up their work, implement the drawings, and inspect and move the product along to the next cell. On the other hand, for example, it is ludicrous to argue that it is reasonable to perform complex test plans from memory.

The next step in our attempt to validate the process document as a viable tier II text is based upon the definition of a quality plan. With reference to ISO 9000:2000, Par. 3.7.5, a quality plan tells you that procedure(s) and resources are required by those who do work, regardless of the type of work that has to be done. Sounds like a process to me. It has inputs (procedures, resources, people), transformations (inputs applied and changed), and outputs (projects, products, processes, contracts). As a result, quality plans are really process documents. The terminology used includes quality-assurance plan, manufacturing control plan, and design control plan. Work orders and travelers are sometimes in the form of a quality plan. This is why we have placed the requirements for plans under the category tier II documents in Table 3.4.

It is also common for the various design phases in the process document to be broken down into subphases. For example, a subphase 1.1 might be the process to create the program plan and a subphase 1.2 might be the process to create the program team. It is not uncommon to nest procedural documents as subphases of the process document. They would still be labeled 1.1, 1.2, ... 1.*N*, but would contain the necessary steps of a tier III document. The idea of nesting appears in the tier III procedure.

It is important to realize that none of the formats are final. They are simply convenient templates to use to increase information flow. That is the sole purpose of the taxonomy—to make the tasks easier to follow and understand. Feel free to innovate to suit your purposes. Tier II and tier III formats are often mixed together. If it works, use it.

The Difference in Format Between Tier II and Tier III Documents At this point in our discussion, it would be best to give an example of the differences in format between our recommended tier II process document/SOP/quality plan and a tier III procedure/work instruction. For this purpose, we will describe an overly simplified design engineering process document for Apogee E&M Inc. (see Table 6.3). The table lacks the usual logo and document control features

Table 6.3
Design Engineering Process for Apogee E&M, Inc.

ISO Par.	Design phase	Activity	Managed by...	Inputs	Outputs	Resources/ Controls	Link to Document
7.3.1	1.0	Design planning	Engineering manager	Marketing requirements document	Program plan Team assignments	Gantt charts Team meetings	Form E-1 Form E-2
7.3.2 7.3.4	2.0	Design inputs	Program manager	Planning documents Regulatory requirements	Specification first draft Design review (DR) Records	DRs	Form E-3 Form E-4 DR procedure
7.3.3	3.0	Design output	Project engineer	Specification first draft	Released specification DR records	DRs	Form E-5 Design review procedure
7.3.5	4.0	Design verification	Project engineer	Released specification Verification test plans	Product verification reports Engineering change notices (ECNs) DR records	DRs ECN board Calibrated test equipment	Form E-6 DR procedure ECN procedure Verification test procedure Metrology procedure
7.3.6	5.0	Design validation	Project engineer	Released specification Product verification reports ECNs Validation test plans	Product validation reports ECNs Acceptance test equipment Bill of material (BOM) Design transfer package DR records	DRs ECN board Calibrated test equipment Customer inputs Marketing and sales inputs Customer-service inputs Operations inputs	Form E-7 DR procedure ECN procedure Validation test procedure Metrology procedure
7.3.6	6.0	Transfer to operations	Program manager	BOM Design transfer package	Release to operations Pilot line testing	Acceptance test equipment and proced-ures	Form E-8 Transfer to operations procedure

such as approvals and dates and the revision level. It could be either a hard-copy or online document. The Standard's paragraphs will also be indicated so that you can see how the Standard can be used as an effective template in the creation of process documents. Most readers will feel that the process document is similar to an SOP or quality plan or control plan, and they are correct.

We will then illustrate a tier III procedure that is linked from the process document to show how self-evident the differences are between the two formats. The tier III procedure will deal with the transfer of product from engineering to operations (see Table 6.4). The location of this procedure in Table 6.3 is highlighted with underlining. Again, the usual document control features are missing for the sake of expediency.

Please note that the format for Table 6.3 is highly idealized to illustrate the classical approach to process documentation. Most writers use the more familiar SOP format. Either way is fine as long as the required information is included (i.e., the design phase, activity, responsible management, inputs, outputs, resources, and controls required), and linkage to pertinent lower tier documents to aid in QMS document navigation.

6.3 ISO 9000 Quality Plans—Optional

The optional requirement for quality plans is stated as a note in ISO 9001:2000, 7.1: Planning of Product Realization, and its definition was discussed previously. About 40 years ago, quality plans were very common in MIL-Q-9858 quality-control systems and consisted of bubble flow charts with all of the associated documentation affixed to the chart. Today, quality plans vary greatly and are an integral part of the QS-9000 requirements [1], and are discussed in some detail in ISO 10005:1995(E) [2].

6.3.1 Sounds Like a Process

A quality plan sure sounds like a process, and indeed it is (i.e., it is a description of a set of interrelated or interacting activities that transform inputs into outputs). As a result, the old bubble chart configuration is as true today as it was 40 years ago and is a very useful rule in the creation of a quality plan graphic (see Figure 6.4).

We have termed the quality plan as optional because the Standard uses the term as a note and notes are informative (guideline) as opposed to normative (required).

Quality or control plans are useful in all types of organizations. Table 6.5 is the first page of a quality plan that could be used for a general contractor.

Table 6.4
Transfer to Operations Procedure for Apogee E&M, Inc.

Transfer Steps	Step Description	Step Activities	Date Planned	Link to Document
1.0	Hold initial ECN transfer to operations review	1.1 Transfer bill of material (BOM) and engineering transfer package to operations 1.2 Transfer acceptance test equipment to operations	1.1 Jan. 02 1.2 Feb. 02	Form E-8 Test equipment operators manual
2.0	Review manufacturing plan	2.1 Review manufacturing engineering manufacturing plan 2.2 Schedule initial pilot run (10 units)	2.1 Mar. 02 2.2 April 02	Form ME-1 Form ME-2
3.0	Run initial pilot line (10 units)	3.1 Monitor and analyze data and prepare NCMRs as required 3.2 Prepare ECNs based on NCMRs as required	3.1 May 02 3.2 May 02	Form ME-3 Form E-6
4.0	Initial manufacturing engineering review	4.1 Review results of initial pilot run 4.2 Review ECNs and take appropriate action 4.3 Close open NCMRs 4.4 Plan for final pilot run (25 units)	4.1 June 02 4.2 June 02 4.3 June 02 4.4 June 02	Records of initial pilot run Records of ECNs Records of NCMRs Form ME-4
5.0	Run final pilot line (25 units)	5.1 Monitor and analyze data and prepare NCMRs as required 5.2 Prepare ECNs based on NCMRs as required	5.1 July 02 5.2 July 02	Form ME-3 Form E-6
6.0	Final manufacturing engineering review	6.1 Review results of final pilot run 6.2 Review ECNs and take appropriate action 6.3 Close open NCMRs 6.4 Schedule final release review	6.1 Aug. 02 6.2 Aug. 02 6.3 Aug. 02 6.4 Aug. 02	Records of final pilot run Records of ECNs Records of NCMRs Form E-8
7.0	Hold Final ECN transfer to operations review	7.1 Transfer final BOM and engineering design package	7.1 Sept. 02	Form E-8

Notice that the documentation requirements are separated into operational and quality-control references.

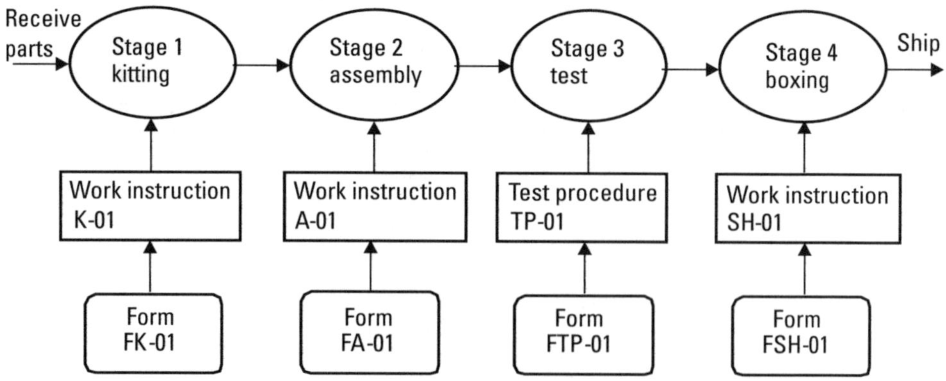

Figure 6.4 Quality plan to build an electronic device. Process flow with documentation = quality plan (when documentation stipulates resources required).

6.3.2 Device Master Record Technique

Another technique used to create a quality plan is to form a device master record (DMR) that either contains or sends the reader to the following [3]:

- Device/system specifications;
- Total manufacturing process specifications;
- Quality-assurance procedures and specifications;
- Packaging specifications;
- Labeling specifications;
- Installation procedures and methods;
- Maintenance procedures and methods;
- Servicing procedures and methods.

It is necessary to make a clear statement in the manual that indeed the quality plan for a given product is formed by the DMR. The DMR is complete when you can prove that the required device can be completely built and shipped to its performance specifications based on only the DMR protocols.

For this medical device protocol, the actual performance of the device throughout its life cycle is captured in the device history record (DHR). The design phases are maintained in the design history file (DHF). The manner in which the device meets its compliance requirements is kept in a technical file. The higher-level documentation is maintained in the quality system record (QSR)—that is, documents not specific to a particular device such as management reviews and metrology procedures. The technique is readily expandable to any organizational product structure and can be termed, for example, the systems master record (SMR), the systems history record (SHR), and so forth.

Table 6.5
Quality Control Plan for the New Construction Project (Sample Page)

Stage	Step	On-Site Activity (Refer to Project Management Procedures)	Operational Documents	Quality and Safety Control Activities	QC Documents/ Standards
I. Job files	1.0	Set up site office files	Project management procedures stage II	Site preparation	Site preparation checklist
	2.0	Superintendent files			
	3.0	Prepare for receipt of contract			
II. Mobilization and start up	1.0	Develop site utilization plan and mobilize temporary facilities	Project management procedures stage IV	Site preparation Post Occupational Health and Safety Association (OHSA) poster Hard hat required sign Location of MSDS sheets	Site preparation checklist Form S.3.4
	2.0	Post required signs			
	3.0	Review building requirements			
	4.0	Establish charge accounts			
	5.0	Arrange for equipment delivery			
	6.0	Establish safety procedures		Hazard communication engineering MSDS sheets Company safety program OSHA form no. 200 First-aid equipment. Protective gear GFI protection requirements	See Appendix S: Safety Reports and Forms in the Project Management Procedure

Table 6.5 (continued)

Stage	Step	**On-Site Activity** (Refer to Project Management Procedures)	Operational Documents	Quality and Safety Control Activities	QC Documents/ Standards
	7.0	Complete site checklist: Company policy Safety program Substance abuse program Hazard communication program	SITEPREP Form R.8		
	8.0	Prepare access to job site	NSC change order	Crossing required	Railroad rules
III Design and engineering	1.0	Subcontractors provide applicable drawings	Final design package	Design package reviewed by engineering	Plans and specs
	2.0	Survey and stake out areas	Plans	Check for accuracy of the layout Field notes	Owner bench marks and baselines are used
	3.0	Locate structures and lines	Plans	Check for accuracy of the layout Field notes	

Note: Based on the work of the Robinson Construction Co., 215 N. West Street, Perryville, MO 63775, General Contractors.

6.4 Process Flow Charts

A number of flow chart software programs are available that are quite capable of clearly defined process flows. The key thing to remember, however, is that a flow chart without reference to associated documents is only half the story. In addition, not all information can be readily placed in a flow chart without obscuring its clarity. As a result, there is always room for supplemental text and complementary tables (e.g., lists of document numbers, forms to be used, and special instructions to the user).

- Primary information—most importantly, if a flow chart is chosen as the means of communication, it should be the primary source of information. If it shares the same data with another document, there is an excellent chance for redundancy and its tendency to confuse the reader.

We have observed flow charts used successfully in both the manual and in lower tier documents. Flow charts are an excellent technique to use to describe both process and the interaction of processes.

- The combination of a supplemental text and a flow chart form the informational document. It's the same document! We have observed that this concept is difficult to grasp. As a result, we have described this issue in Figure 6.5. The details of a typical flow chart are shown in Figure 6.6. Notice the use of documentation references in both of the examples.

Imports There is a caution on flow charts with regard to computer files for an online system. In some cases, the files for the supplemental text and the files for the flow charts cannot be readily integrated due to import issues. This complicates the search function somewhat but is resolvable with training. It is best to check out the import characteristics of the software so that the ease of flow chart usage and its inherent clarity is not invalidated. (Refer to Appendix D for an example of the text-plus-flowchart-equals-process document.)

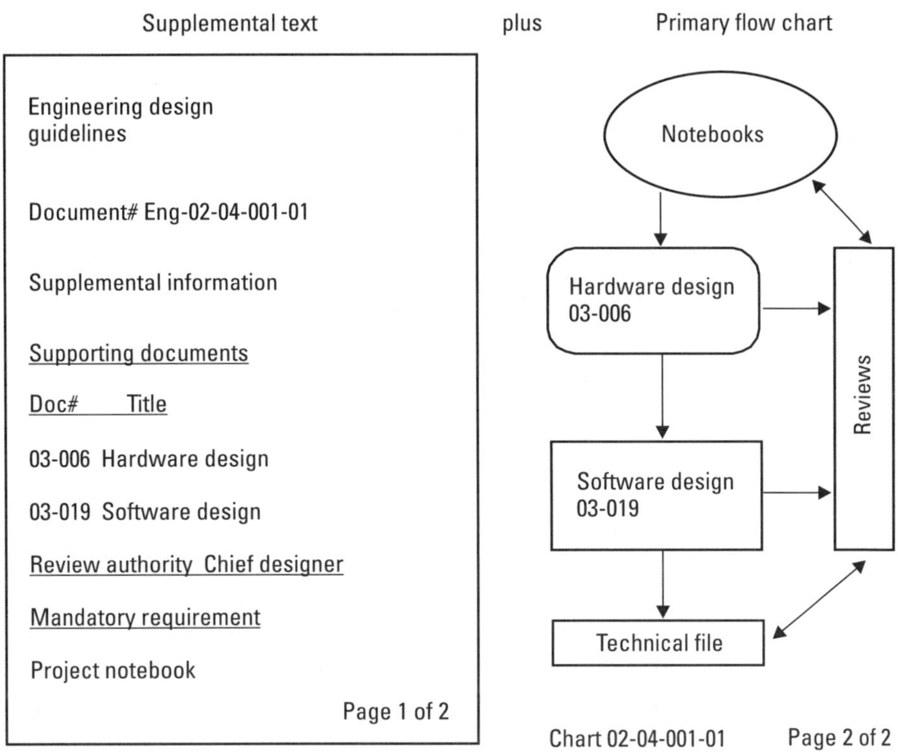

Figure 6.5 A typical flow-charted process structure.

Figure 6.6 Example of a flow-charted process.

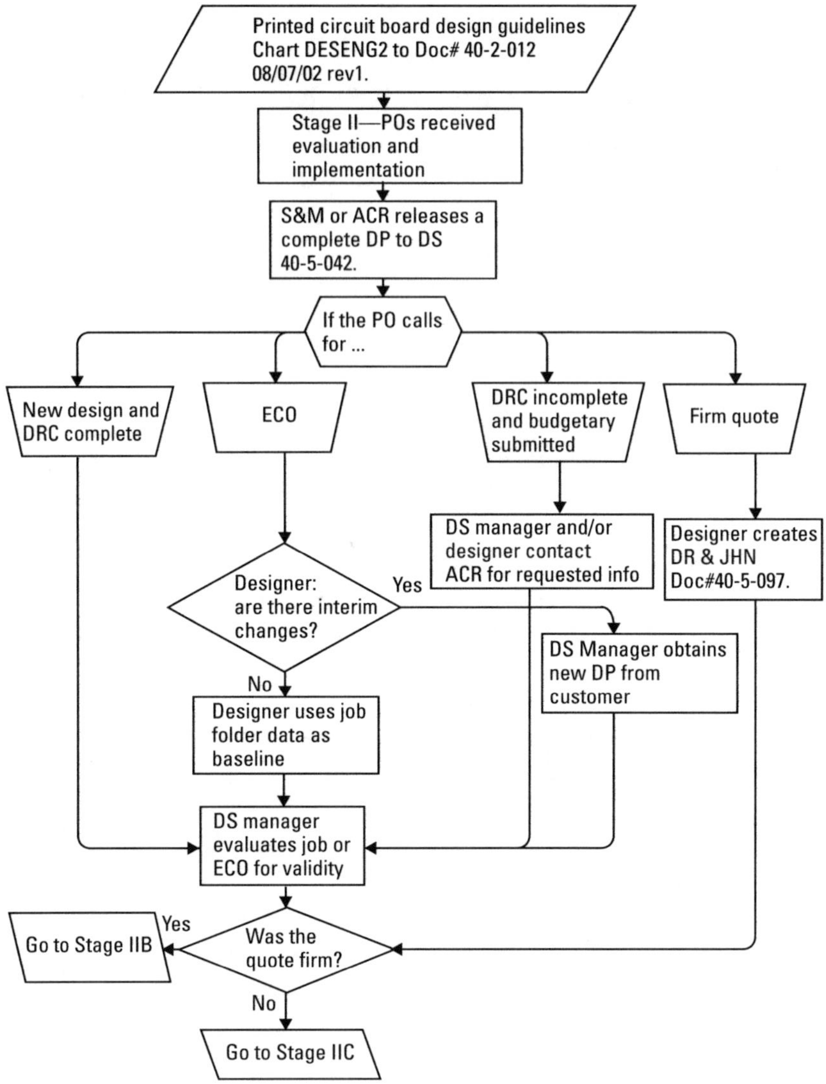

Endnotes

[1] See, for example, Advanced Product Quality Planning (APQP), Reference Manual, from AIAG, June 1994, tel: (810) 358–3003.

[2] ISO 10005:1995(E) presents several typical quality plan configurations that include plans for service organizations, manufactured product, processed material, and a software life cycle.

[3] 21 CFR Part 820 Quality System Regulation, FDA/CGMP Sec. 820.181, October 7, 1996.

CHAPTER

7

Contents

Procedure Design

7.1 Some Procedures Are Mandatory Documents

Par. 4.2.1(c) of the Standard leaves no doubt that there are certain documented procedures that are clearly required as part of the QMS. However, only six are stipulated:

1. Control of documents (Par. 4.2.3);
2. Control of records (Par. 4.2.4);
3. Internal audit (Par. 8.2.2);
4. Control of nonconforming product (Par. 8.3);
5. Corrective action (Par. 8.5.2);
6. Preventive action (Par. 8.5.3).

The rest is up to you if you feel that more procedures are necessary. However, from our analysis to this point, it is clear that more documented procedures are needed just to put the tier II requirements somewhere (e.g., where do we put the plans that define our processes?). In fact, as we have already noted, if we want to use process documents, they can be termed procedures under the broad definition provided in the Standard's vocabulary and guidelines. Table 6.4, discussed earlier, is an example of a "you do this, then you do this" procedural structure that fits our definition of procedure as illustrated in the four-tier taxonomy.

For those writers who are required to include sector-specific requirements (e.g., medical device requirements), be aware that the tendency is in the direction of not only adopting the ISO 9001:2000 process structure but to continue to expand clauses to cover sector-specific mandatory conditions that include more procedures like the 1994 version.

7.2 The Special Case of Work Instructions—Optional

Par. 7.5.1(b) of the Standard conditionally calls for work instructions as part of the QMS-controlled conditions. They are to be available as applicable and as necessary. In this way, not only is extensive discretion afforded to the QMS designer, but the requirement does not require documented work instructions. In fact, the term work instruction is not defined in the canonical texts. So why even mention the term?

The answer to this question is lost in production antiquity. Because I am somewhat ancient, I can recall the term in use over 45 years ago. Clearly, it is locked into the work center concept and the fact that we do work to create a successful enterprise. Work instructions were included in the first released Standard in 1987. In that case, it was a SHALL and required documented work instructions [1]. Work instructions were also noted as supplemental documents in MIL-Q-9858A, Par. 4.1, 1963, and the supplier was responsible for their control. I actually worked with these documents.

So, to make a long story short, Table 6.4 is a perfectly good work instruction. It is the type of document that is kept at the workbench level and used in day-to-day operations. Work instructions are procedures in the fullest sense of the word, as we have defined procedure within our tier III documentation hierarchy. They need not be documented but often are and form a powerful basis for a higher-tier documentation.

Typical work instructions sound like ESD operator-check work instruction, order entry work instruction, screen printer work instruction, subassembly test work instruction, vision monitors installation and configuration instructions, and tape changing instructions. The document is designed for and generally used by an individual worker, in a work center, working on a very specific task.

Work instructions can also be found embedded within a high-level procedure. An example of this format is shown in Table 7.1. In Table 7.1 we show an abbreviated three-phase front-end process for a company that receives a certificate of analysis with some of their raw material (e.g., steel rods). The process flow is from receiving to inventory to kitting. Each part of the process flow has several activities.

The several activities have a number of work instructions shown in the far right column. The forms required are also noted, as well as specific responsibility for the activity. An integrated format of this type is very powerful in terms of simplicity and necessary detail.

Table 7.1
Integrated Process and Work Instruction Document

Process Phase	Activity Description	Work Instructions
1.0 Receiving	1.1 Receiving—at dock (receiver)	1.1.1 Log in the material (F-1.1.1) 1.1.2 Check for the certificate of analysis (C of A) 1.1.3 Send C of A to QA
	1.2 Receiving inspection (QA)	1.2.1 Compare material properties to C of A and log in (F-1.2.1) 1.2.2 Use RTV form if not acceptable (F-RTV-1)
2.0 Inventory	2.1 Inventory—production control	2.1.1 Enter material into the MRP system—note C of A received 2.1.2 Release modified schedules to production
	2.2 Materials movement—coordinator	2.2.1 Move C of A material to special storage 2.2.2 Log in (F-2.2.2)
3.0 Kitting	3.1 Kitting—materials control	3.1.1 Check to see that C of A was received 3.1.2 Obtain pick list 3.1.3 Compare to BOM
	3.2 Kit release—coordinator	3.2.1 Final inspection of kit (F-3.2.1) 3.2.2 Attach traveler (F-3.2.2) 3.2.3 Release kit to assembly

Endnote

[1] Par. 4.9.1 of ISO 9001:1987.

CHAPTER

8

Contents

Forms and the Control of Records

8.1 Forms Versus Records

8.1.1 Formats

Forms/formats represent tier IV of the documentation pyramid. They are essentially templates within which data/information is deposited for analytical or informational purposes. When a form is filled in with data and signed and dated by the observer and filed away for a specific period, it is termed a *record*. The format is controlled by any number of methods. The most common control method is to assign a control number to the form. For example, a form used for internal audit reports might be labeled, F-8.2.2-002, with a revision level, Rev. 3. The form is maintained within the engineering change order (ECO) or document change order (DCO) system used to control higher tier documents.

It is also customary to include more complex formats, such as schematics and assembly prints, into tier IV, although such documents, when they are archived, are actually records because they are signed off and dated. Because the documents contain information and usually state specific work instructions, they could just as easily be classified as tier III documents. Consistency is more important than which tier is chosen for a specific document. The development of a master forms list is the most important element so that users can quickly find an appropriate form, and effective audits can be performed.

8.1.2 Analytical Linkage

Records are basically historical documents and are not part of the operational linkage. They are certainly part of the analytical linkage that forms an information context in parallel with the operational process. The Standard does realize this fine distinction in 4.2.3: Control of Documents and reminds us that records are to be controlled and that they form a unique category of documentation. However, invariably, records are shown as part of tier IV. I remain ambivalent on the subject, and my efforts in an audit are primarily spent to ensure that the records are clearly defined, in good shape, and under control.

The difference between operational and analytical linkage can be demonstrated if one considers that during the manufacturing cycle, the operator refers to a work instruction—a dynamic document. The operator does not refer to all of the process data sheets signed off over the past month to find out what to do in the next step.

However, the quality-assurance manager analyzes all of the process data sheets over the past month to look for trends in nonconformances and time-related issues.

The operator fills in the process data sheet form—a tier IV document—and the quality-assurance manager creates a nonconformance Pareto chart from the process data sheets. This becomes a record of nonconformance frequency versus months and years past.

It can get tricky. For example, a marketing and sales forecast of potential received orders (bookings) is a record that captures the marketing and sales estimate of anticipated bookings on a particular date. The record of this estimate establishes booking goals. This time-based set of goals becomes one of the data lines on the run chart of bookings obtained versus fiscal months. The next line on the run chart would be a record of actual bookings plotted in time. The combination of the two lines on the chart constitute a record of actual bookings versus anticipated bookings.

In contrast, a corrective action procedure is not a record. It is a dynamic document. However, the previous revision of the procedure is a record if it is kept because of legal or regulatory requirements (e.g., for the FDA). Sometimes, we find the current revisions of policies, processes, procedures, and work instructions included on a master records list. There is no harm in this activity as long as the auditee understands the difference between document and record control and has established the proper procedures to carry out such protocols.

This difference between forms and records is illustrated in Figure 8.1 and Table 8.1.

8.1.3 Bypasses for Forms

It is unnecessary to create documents so that every tier is covered—do what makes sense. In auditing, I live by the rule: If it sounds stupid, it's stupid.

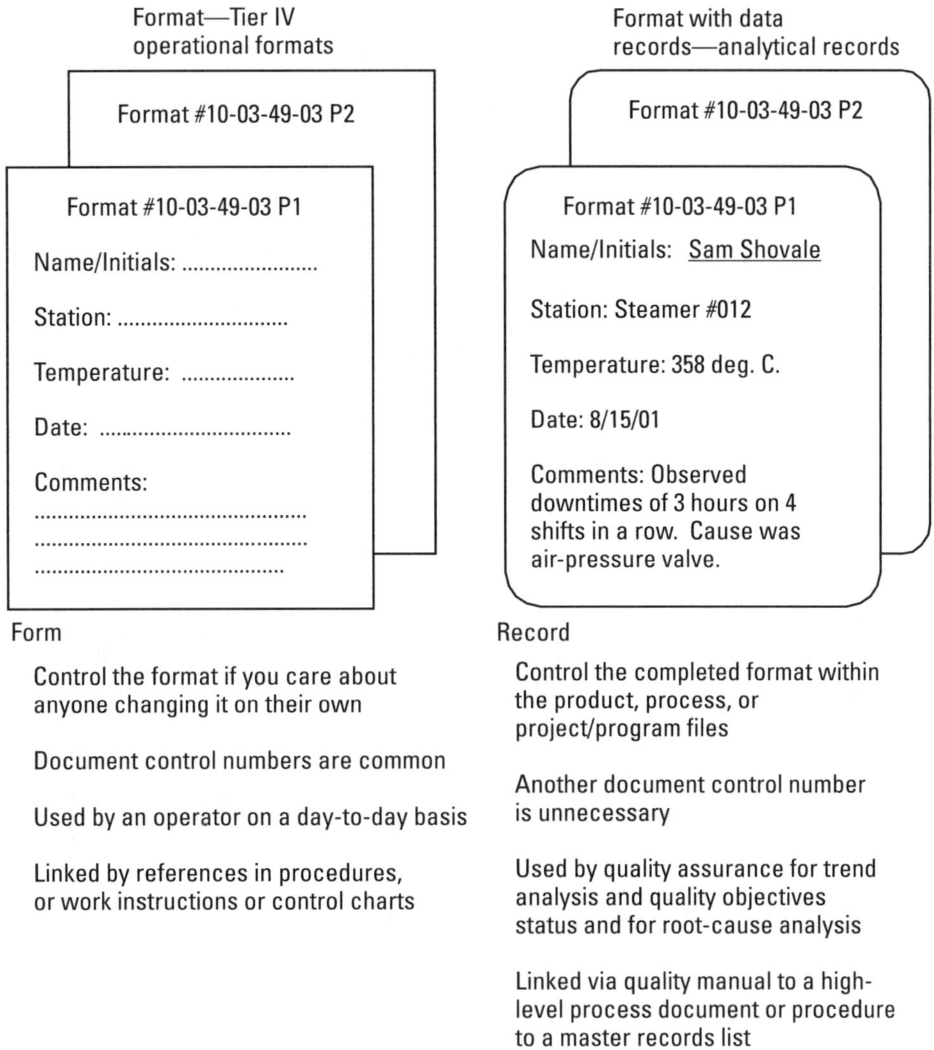

Figure 8.1 Operational versus analytical linkage.

Sometimes I hear my clients murmuring to themselves, "sounds stupid—let's try something else," so it seems like a pretty good rule. Works every time. In other words, if it is required to go from the manual to a form (e.g., the format for management reviews), do it. It is not necessary to create an SOP and a work instruction to get to that form.

Figure 8.2 illustrates this concept. In the left-hand-side drawing, the forms are directly linked from both a procedure and a work instruction. In the middle drawing, the forms are directly linked from a work instruction. In the right-hand-side drawing, the forms are directly linked from the quality manual.

Table 8.1
Forms Versus Records

Format—Tier IV Operational Formats	Format with Data Records—Analytical Documents
Form	*Record*
Control the format if you care about anyone changing it on their own	Control the completed format within the product, process, or project/program files
Document control numbers are common	Another document control number is unnecessary
Used by an operator on a day-to-day basis	Used by quality assurance for trend analysis and quality objectives status and for root cause analysis
Linked by references in procedures, work instructions, or control charts	Linked via quality manual to a high-level process document or procedure to a master records list

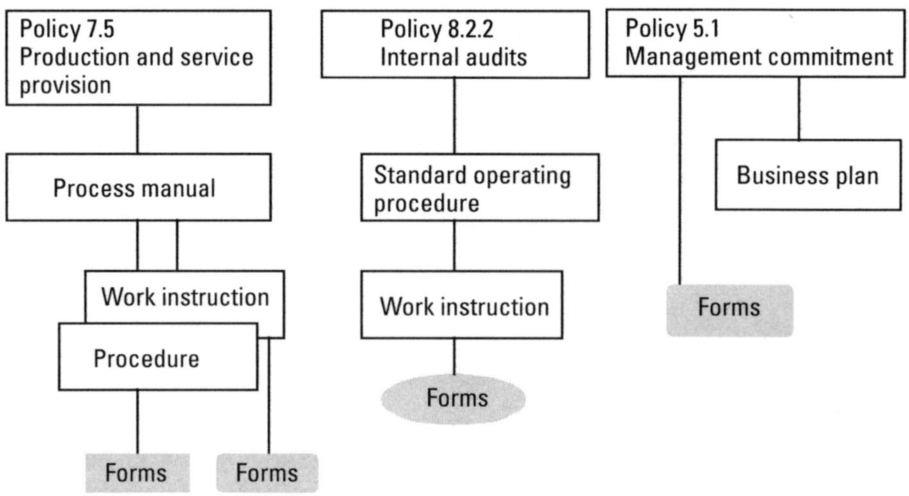

Figure 8.2 Possible ISO 9001:2000 forms linkage schemes.

8.2 Records Are Mandatory Documents

Par. 4.2.1(e) of the Standard is very clear about records as mandatory documents. Appendix B of the *ISO Guidance on the Documentation Requirements of ISO 9001:2000*, provides a list of such records (see Table 8.2.) Notice that the required records are given in a descriptive manner, and it is up to the organization to clearly define which specific records are to be kept. Typical records of this type are indicated in Table 8.2.

Table 8.2
Records Required by the Standard with Typical Actual Records Maintained

Clause	Mandatory Record	Typical Records
5.6.1	Management Reviews	Minutes of the monthly review Weekly departmental reviews In-house publications
6.2.2(e)	Education, training, skills, and experience	Resumes Certificates of course work (internal/external) External studies/seminars Individual employee training record Signed off training attendance sheets Annual reviews Syllabus for training session On-the-job (OJT) training records Training schedules and plans
7.1(d)	Evidence that the realization processes and resulting product fulfill requirements	Nonconforming material reports (NCMRs) Returned material authorizations (RMAs/RGAs) Repair history sheets Routers/travelers Supplier certificates of analysis (C of Cs/C of As) Inspection and test stamps Waivers/concession reports Customer release reports Shelf-life records ESD records Shipping records Customer site reports Warranty records
7.2.2	Results of the review of requirements relating to the product and actions arising from the review	Requests for quotes, proposals Catalog quotes Purchase orders Sales orders Sales acknowledgments
7.3.2	Design and development inputs	Market analysis Regulatory standards Requirements document Competitive product analysis Research and development results
7.3.3	Design and development output—not explicitly called out as requiring records; these are "implied" records	Software functional specification Hardware functional specification User's manual Risk and hazard analysis Systems specification Verification and validation test protocols Bill of materials

Table 8.2 (continued)

Clause	Mandatory Record	Typical Records
7.3.4	Results of design and development reviews and any necessary actions	Design review minutes Final drawings Final schematics Final systems specification Transfer to production plan
7.3.5	Results of design and development verification and any necessary actions	Test data packages Alpha testing results Final design modifications Verification and validation of software completed Summary verification test report Controlled laboratory notebooks verified
7.3.6	Results of design and development validation and any necessary actions	Test data packages Beta testing results Customer validation reports Summary validation test report
7.3.7	Results of design and development changes and any necessary actions	Engineering change orders Engineering change requests Revised drawings and prints Transfer to production sign off
7.4.1	Results of supplier evaluations and actions arising from the evaluations	Approved supplier list Evaluation of supplier performance Supplier corrective action reports Certificates of compliance/analysis Source inspection reports Incoming receiving reports Surveys of suppliers Supplier audit reports
7.5.2(d)	As required by the organization to demonstrate the validation of processes, where the resulting output cannot be verified by subsequent monitoring or measurement	Computer validation report Customer numeric control (CNC) validation report Employee qualification report Process validation report Machine validation report Revalidation reports Employee requalification reports
7.5.3	The unique identification of the product, where traceability is a requirement	Serial number logs Control of stamps logs Receiving logs Inspection reports Testing reports Heat numbers Certificates of compliance/analysis Pick lists Packing lists

Table 8.2 (continued)

Clause	Mandatory Record	Typical Records
7.5.4	Customer property that is lost, damaged, or otherwise found to be unsuitable for use	Nonconformance material reports Nonconformance product reports Returned material authorizations Returned goods authorization Product repair reports
7.6(a)	Standards used for calibration or verification of measuring equipment where no international or national measuring standards exist	In-house designed test fixtures Software verification and validation reports Ad hoc industrial Standards In-house designed measuring fixtures Golden electronic boards
7.6	Validity of previous results when measuring equipment is found not to conform with its requirements	Corrective action reports Nonconforming product reports Recall reports Returned goods authorizations
7.6	Results of calibration and verification of measuring equipment	Test hardware validation Test software validation Inspection equipment calibration Measuring equipment calibration Test equipment calibration Master calibration lists
8.2.2	Internal audit results	Audit schedules with assignments Audit plans and checklists Audit reports Vendor/supplier audit reports Third-party audit reports Customer audit reports
8.2.4	Evidence of product conformity with the acceptance criteria and indication of the authority responsible for the release of the product	Receiving inspection and test reports In-process inspection and test reports Final inspection and test reports Vendor/supplier certificates of compliance/analysis Your certificates of compliance Declaration of conformity Pass/fail records Quality-control stamps Quality-assurance release records Device history records
8.3	Nature of the product nonconformities and any subsequent actions taken, including concessions obtained	Rework records Scrap records Nonconformance tags Hold tags Customer release records Waiver records Concession reports

Table 8.2 (continued)

Clause	Mandatory Record	Typical Records
8.5.2	Results of corrective action	Corrective action reports (CARs) Supplier corrective action reports (SCARs) Audit corrective action reports (ACARs) Registrar's nonconformance reports (NCRs) Summary presentation at the management review
8.5.3	Results of preventive action	Preventive action reports (PARs) Root-cause analysis reports Action team reports Summary presentation at the management review

This area tends to be one of mass confusion due to a lack of specificity [1]. Records (used as objective evidence of activities) complement the hierarchal documents and can be associated with any tier. For example, records do not require a separate documentation control numbering system because they are already controlled, either centrally or locally, by date and signature.

Imagine how confusing it would be to take a form with a control number F-103-01, fill it in, and then give it another control number R-103-01 for storage as a record. This is a danger in ISO document control interpretation and is certainly not specified in the Standard.

Another case of confusion can occur when forms are signed off to approve their distribution. In that case, it really gets confusing between the form's approval signature and the signature of the operator who signs off in the data columns. Approval sign-offs on forms should be avoided and are usually removed once the issue is discussed.

8.2.1 Records As Historical Documents

A record is basically an historical document that contains information that is worth keeping for some time. The most familiar form of record keeping is the documents we maintain for the Internal Revenue Service. For an FDA-regulated organization, the need to maintain device history records is made painfully clear via U.S. government penalties [2]. Records are usually filled-in forms, but they can be in the form of memoranda, reports, or e-mails.

8.2.2 Records As Objective Evidence

The Standard's vocabulary requires that a record should contain useful information that either lists achieved results or provides evidence that some operational activity was performed. This set of requirements provides the framework for an expansive list of specific records (see Table 8.2).

To create a meaningful set of records requires that we do not use just the category sales and marketing records, but that we define sales and marketing records explicitly (e.g., quotations, purchase orders, sales orders, acknowledgments, purchase order changes). We also specify where they are kept (e.g., maintained in the sales files cabinet). Further, we specify who maintains the records (e.g., maintenance of the records is by the sales and marketing administrative supervisor). In addition, we specify retention time (e.g., all records are kept for the current year plus 2 years). Finally, we specify who can destroy records (e.g., records cannot be destroyed without the direct approval of the controller). It is usually best to avoid specifying the exact nature of record disposal and state that it is at the discretion of top management (e.g., the controller).

8.3 The Records Master List

An excellent rule that can be used to form the records master list is to first list all of the forms used by the organization. Invariably, most of those forms will be kept by someone in their files as a record of their acceptance, rejection, verification, identification, and categorization of their activities. Then, add in the more subtle records, such as management reviews, design reviews, responses to requests for proposals and quotes, and preventive action reviews.

8.3.1 Specific Records

As discussed previously, the Standard does call out a specific number of required records. However, as we saw, the classification is generic. As a result, it was necessary to interpret the spirit of the requirement. The shock on the faces of the records steward when the assessor asks to see a list of documents contained in the contract review files is not necessary.

Just remember that an assessor cannot assume what belongs in those files. The assessor must audit against what has been declared by the supplier as a quality record. The declarations can be challenged if they do not comply with the Standard or it is discovered that many documents are kept as records but are not included in the master records list. Organizations often keep more records than are specified in the Standard, and such records need to be listed because they are an integral part of the QMS.

Table 8.3 illustrates how a single chart can be used to clearly define organizational records. The chart indicates the key parameters of process recorded, type, responsibility, location, and retention time.

Notice that the category "master list" is included. Such lists are normally locally controlled documents that require a name and a date. However, they

Table 8.3
Partial Master Records List of Typical Records

Business Process	Record(s) Description (Form Used—A Few Samples)	Retained by (Primary Copy)...	Location [Online or Hard Copy (HC)]	Retention Time
Executive	Quarterly management review minutes (FE-001)	ISO 9000 management representative	Online	Current plus one (*C* + 1) year
	Business plan (FE-002)	President	Online	*C* + 3 years
	Quality policy (FE-003)	ISO 9000 management representative	HC (posted)	Current version
	In-house publications	Executive assistant	HC	*C* + 1 year
	Market analysis	Vice president of sales and marketing	Online	*C* + 3 years
Marketing and sales	Quotes (FS-001)	Sales administrator	HC	*C* + 5 years
	RFQs and RFPs		HC	*C* + 5 years
	Catalog price lists (FS-002)		Online	Current version
	Sales orders (FS-003)		HC	*C* + 5 years
	Sales Acknowledgments		HC	*C* + 5 years
	Purchase orders		HC	*C* + 5 years
Engineering	Engineering change requests and notices	Engineering document control manager	HC	5 years after life of device
	List of Standards		Online	
	Technical files		HC	
	Design review minutes		Online	
	Risk and hazard analysis		Online	
	Device master record		HC	
Operations	Preventive maintenance reports	Plant engineer	HC	*C* + 2 years
	Shelf life records	Warehouse manager	HC	*C* + 2 years
	ESD records	Production manager	HC	*C* + 1 year
	Receiving records	Warehouse manager	HC	*C* + 3 years
	Training records	Vice president of operations	Online	End of employment
	Purchase orders	Purchasing manager	HC	*C* + 3 years
	Approved vendor List	Purchasing manger	Online	*C* + 1 year
Quality assurance (QA) and regulatory affairs	Nonconforming material reports	QA document control supervisor	HC	5 years after life of device
	Device history record		HC	
	Supplier certificates of compliance and analysis		HC	*C* + 1 year
	Corrective action reports		Online	*C* + 1 year
	Preventive action reports		Online	*C* + 1 year
	Master document lists		Online	*C* + 1 year
Servicing	Return material authorizations	Administrative assistant	HC	*C* + 1 year
	Service reports		HC	*C* + 3 years

also contain information (e.g., current revision level, the location of the document, and who it is signed out to) that the assessor requires for objective evidence of, in this case, document control. This puts them into the category of a record.

8.3.2 Records Quantity

The number of records maintained by an organization is always a "bone of contention" with top management. Unfortunately, the harder you push to remove records, the more you will find. People need them to do their jobs—just make sure they are useful. I have witnessed an exercise in which an executive committee sat down to decrease the number of forms used by the company. Out of about 130 forms they removed nine ... but added 16 more.

Scrap paper and computer transfer notes are not records. It is important to know when a form is maintained for a significant reason because many forms are really scrap paper and need not be controlled.

A good example of this is in a basically online system where raw data is taken daily and fed into a computer database at the end of the day. The form used to collect the raw data could have been just a notebook page, and, as a result, raw data sheets can be discarded. In this case, the record is controlled on electronic media.

Endnotes

[1] The subject has received some deserved dedicated attention. See, for example, Brumm, Eugenia K., "Managing Records for ISO 9000 Compliance," *Quality Progress*, January 1995, p. 73. The book of the same title is also available from the ASQ, ISBN 0-87389-312-3.

[2] Quality System Regulation, Part VII, Department of Health and Human Services, Food and Drug Administration, 21 CFR Part 820, Current Good Manufacturing Practices (CGMP): Final Rule, October 7, 1996, published by the SAM Group, Stat-A-Matrix, Edison, N.J., Sec. 820.181: Device Master Record.

CHAPTER

9

Contents

Other Mandatory Documents

9.1 SHALL Analysis of Other Mandatory Documents

To borrow a phrase, "Discretion is the better part of the QMS." It's everywhere! Consider the requirement Par. 4.2.1(d), which alerts us to the fact that we need to produce documents that result in effective process planning, operation, and control. A SHALL of this type must be broken down into a little "SHALL analysis." There are three mandates to this one liner, and we will use a matrix to clarify this imperative (see Table 9.1) [1].

Those who believe that only a few documents are really needed for the new Standard are well advised to take a cautious position on this matter because of Par. 4.2.1(d) of the Standard. For example, the perfect auditor question is "which documents do you use to ensure the effective planning, operation, and control of your processes?" In this case, "effectiveness" is defined (i.e., Par. 3.2.14 of the vocabulary) and we are informed that we are effective when what we plan to do gets done.

Clearly, it is best to have a well-documented set of tier I, II, III, and IV documents. Of course, we contend that this will produce an overall effective QMS because all of the requirements of the Standard will be covered. Thus, a win-win scenario. A checklist to help organize your thoughts in this matter is addressed in Appendix C [2].

Table 9.1
Other Mandatory Documents "Shall Analysis"

Requirement	Key Word	Fits into This Tier with Typical Documents Called...
The documents required to ensure effective process planning	Planning	Tier II: Quality plan Control plan Business plan Manufacturing plan Manufacturing engineering plan Test plan Quality-assurance plan Technology transfer plan
The documents required to ensure effective process operation	Operation	Tier II: SOPs Process maps Process flow charts Work orders Routers Tier III: Work instructions Process instructions Functional test procedures
The documents required to ensure effective process control	Control	Tier III: SPC charts Pareto charts FMEAs Risk-analysis reports Yield reports Tier IV: All of the formats Templates Drawings Schematics Check sheets used to collect data for analysis and record keeping

9.2 The Special Case of Product Characteristics

We have now reached the last requirement summarized in Table 3.4 (i.e., the discussion of Par. 7.5.1, of the Standard), which deals with describing the characteristics of product. We are to provide information for this purpose but are not told in what form it is to occur. The information need not be documented. If we state that it makes sense to prepare such information in documented form, we create a specification and the ISO 9000:2000 vocabulary informs us that specifications are documents that address requirements (Par. 3.7.3). So we can solve the Catch 22 problem if we appeal to reason. In our hierarchal documentation structure, we suggest the use of documented specifications, marketing requirements, and technical brochures and manuals to accomplish this purpose. We just cannot see how verbal information can be stable enough to work in a real enterprise scenario.

According to the vocabulary, the specific characteristics are essentially unlimited in that they can be either qualitative or quantitative and include all manner of categories such as physical, sensory, behavioral, temporal, ergonomic, and functional.

However, we normally speak in terms of preliminary, final, critical, intermediary, regulatory, and validated specifications to describe a product with regard to form, fit, function, performance, safety, and environmental behavior. The documents include specifications, drawings, schematics, test plans, blueprints, work orders, and travelers and routers. As a result, the documents needed fit into all three of the lower level tiers (i.e., II, III, and IV).

For example, a marketing brochure that contains explicit product specifications should be a controlled document at the tier II level. A test plan that includes specific performance values and tolerances should be a controlled tier III document. A drawing under engineering change-order control is normally classified as a tier IV document.

Critical characteristics are also highlighted on drawings and are a key ingredient of conditions set down by automotive companies engaged in QS-9000 certification protocols. A typical term used would be Control Item (∇) Parts, where the inverted delta is used to denote the parts that have critical characteristics (e.g., dimensions or performance tests that could affect safe vehicle operation or impact compliance with regulatory requirements) [3].

9.3 Mandatory Organizational Requirements

9.3.1 Mandatory Requirements from the Registrar

The registrar requires that we define the employee certification scope because this is one of the parameters used to calculate how many assessors are required for how many days to effectively complete a certification assessment. The assessors need to know exactly how the personnel are distributed over what areas of the facility as a function of departmental activities. A common and effective way to do this task is by means of an organizational chart or an equivalent table. A chart is generally used and includes a box for the ISO management representative who normally reports directly to the site manager for ISO 9000 purposes.

9.3.2 Responsibility and Authority Required by the Standard

Clause 5.5.1 of the Standard mandates that we are to clearly define responsibility and authority within the organization and propagate such information throughout the organization.

A typical organizational chart is shown in Figure 9.1. Notice that all levels of the organization are defined. Quite often, the charts are found as an appendix to the main body of the manual. The actual names of employees are not required, although assessors are very grateful to have such a chart available in addition to the generic one.

It is also important to include a paragraph description of the duties and responsibilities for each of the top managers. This could be an appendix but is normally placed in the body of 5.5.1: Management Responsibility.

We also see in Figure 9.1 that the various activities implied in Par. 5.5.1 of the standard are included in terms of a code in each appropriate box. The code is explained in Figure 9.2. In the 1994 version, this area of responsibility and authority was a strong area of contention as to why this requires a detailed response. The new version is less prescriptive, but just as powerful because it clearly addresses all processes and does not emphasize manufacturing-type activities, as it did in the 1994 version.

A typical paragraph would read as follows:

Engineering manager's responsibilities:

1. Assists marketing and sales in sales of new products;
2. Supports preparation of the marketing requirements document;
3. Obtains and allocates engineering resources;
4. Serves as the chief engineer;
5. Schedules and supervises engineering projects;
6. Ensures compliance to ISO 9001 design requirements;
7. Manages the transfer of product to manufacturing;
8. Supports manufacturing with continuing engineering;
9. Interfaces with customers to determine design satisfaction.

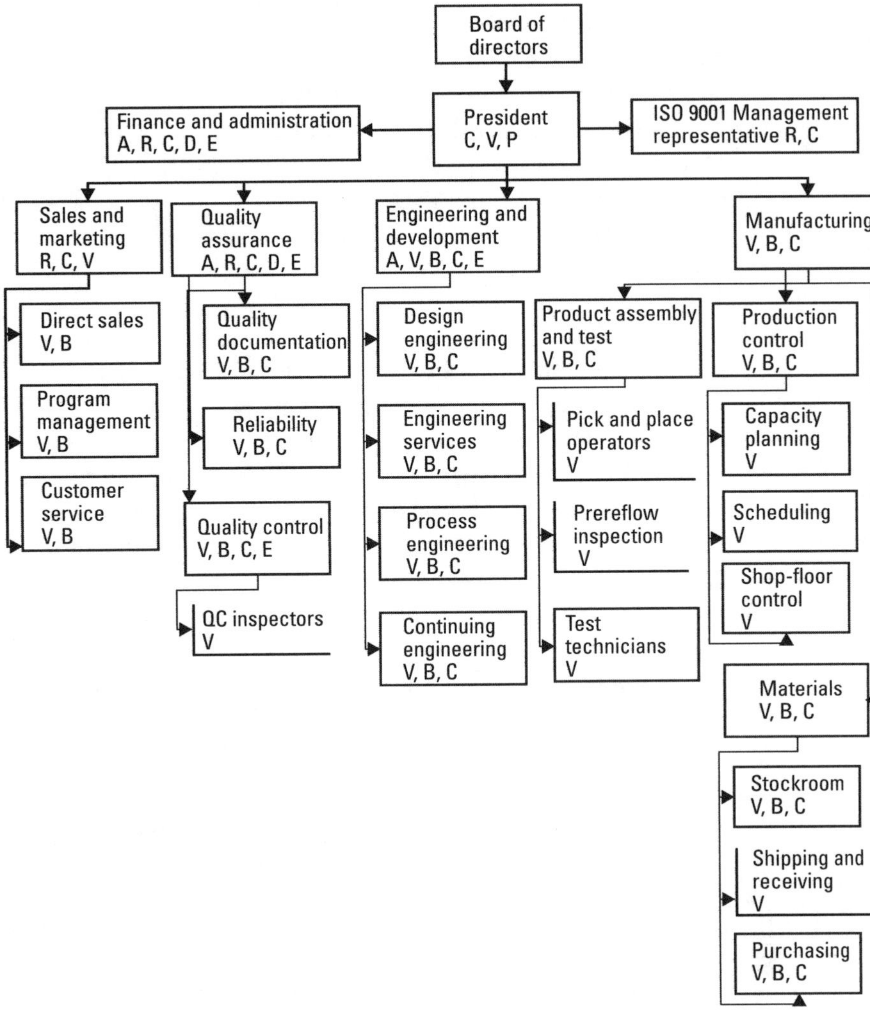

Figure 9.1 Excellent Corporation's organization.

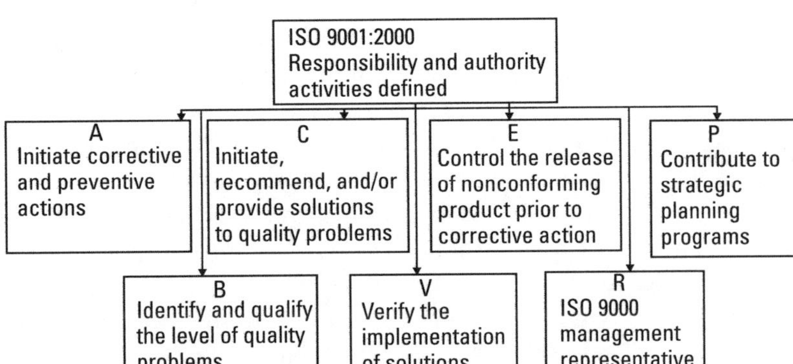

Figure 9.2 Responsibility and authority legend for Excellent's organization.

9.3.3 Job Descriptions

Job descriptions can also be used to define these functions, and specific steps in procedures can be used to enhance the descriptions. However, the manual must respond to this SHALL and be clear as to the method used to define and communicate such activities. Job descriptions should clearly indicate the requirements for education, necessary skills, acquired training, and related experience to the degree required for a given employee position or title.

9.3.4 Registrar Mandatory Interface Issues

In many cases, the certification site is part of a much larger organization. It is then necessary to define the interfaces that exist between the corporate offices and interdivisional sites. This somewhat obscure requirement is extremely important in multidivisional organizations that share operational areas (e.g., engineering, purchasing, metrology, and shipping and receiving) [4].

A typical example of such an interface chart is shown in Figure 9.3. As indicated, there are a number of corporate and divisional interfaces. The same information could be demonstrated by means of a table. In fact, when there are several dozen interfaces, a table is easier to understand.

What we see in Figure 9.3 is that corporate marketing and sales provides market analysis and customer leads to the general manager. The general manager works with the in-house customer service department to close and follow up on sales. In addition, the local finance department interfaces with the corporate controller for capital equipment and fiscal budget planning.

Also, the local design capability is enhanced via the research and development facilities at the corporate level, and local purchasing obtains better price discounts by buying raw material through the corporate materials department.

In many situations, the purchase of material from another division is by means of the same purchase order that is used to buy from any subcontractor. Also, when product is shipped from one division to another the same sales orders are used as with any shipment to a customer. In that case, it is usually acceptable to state in the manual in response to 4.1: General Requirements that "All buy and sell transactions between the Excellent Corporation's operating divisions are by means of either purchase orders or sales orders that are the same formats used with either vendors or customers."

This type of blanket statement fulfills both the requirements to identify processes and to control outsourced material.

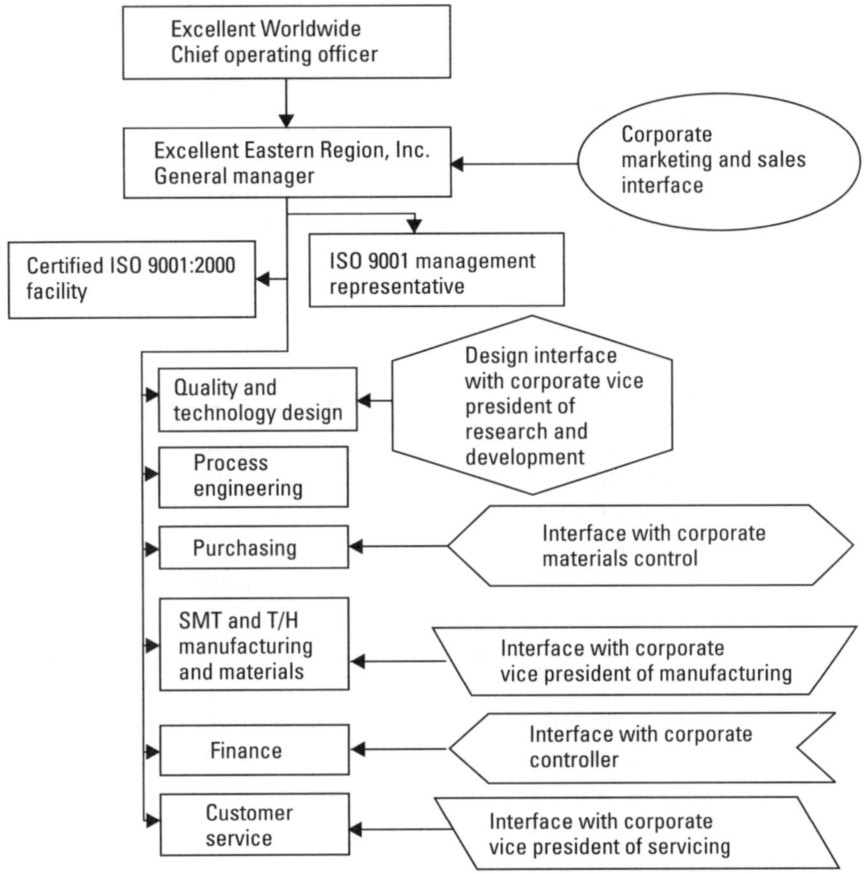

Figure 9.3 Excellent Corporation's interdivisional relationships.

Otherwise a fairly detailed interdivisional agreement would need to be created—usually in the form of a procedure—and signed off by the two interfaced general managers. Be sure to pass such decisions by your registrar for their acceptance and comments.

A chart of this form is also extremely useful for those organizations that primarily use subcontractors to design and manufacture their products. In that case, the subcontractors may fulfill many of the roles otherwise supplied by a corporate division.

9.4 Mandatory Effective Implementation Requirement

The requirement for effective implementation is expressed in Par. 4.1: General Requirements, in prescriptive form in Par. 8.4: Analysis of Data, and in more prescriptive terms in Par. 8.5.1: Continual Improvement.

This use of redundancy (previously classified as concomitance) offers a great deal of help to those who had trouble with the 1994 text in that it clearly defines the basis for metrics to establish QMS effectiveness. For example, its descriptive language on audits now encompasses all audit categories—conformance, compliance, system, process, product, internal, vendor, and customer. Additionally, design reviews and customer complaints are also necessary parameters upon which to base effectiveness—we need to also include these two categories into our analytical studies.

9.5 Nonmandatory Sensible Requirements

The charge to the authors to create a reasonable volume of documents and to keep the corporate economics in mind is expressed in Par. 4.2.1: General of the Standard as note 2 and is therefore not mandatory but is to be considered a guideline [5]. We are alerted that QMS documentation can differ widely between organizations as a result of such characteristics as size, organizational structure, process complexity, and levels of personnel training, skills, experience, and education. The similar clause in the 1994 version was mandatory.

At issue here is the tendency to overwrite—usually a good 40% more than is necessary for an effective presentation. We all do it, and it usually takes a few years after certification to streamline the system. Typical forms of redundancy include the following:

- Policy statements in the quality manual repeated in the tier II and tier III documents;
- Tier III documents that repeat the same procedures as the tier II documents;
- Flow charts with associated text pages that state the same information;
- Master lists repeated in labeled text in procedures such as a master records list with a list of records repeated in a procedure. One or the other is sufficient as long as all the necessary record requirements are met.

There will be plenty of time available after the ISO 9001:2000 certificate hangs on the wall to optimize the documentation. Besides, the documents will most likely have a number of significant changes during the first surveillance period, so that is a good time to make the necessary revisions.

9.6 Special Mandatory Requirements

9.6.1 Customer Complaints As a Mandatory Requirement

In its charge to the accredited registrar, EN 45012, Par. 18 advises us that the registrar is to make sure that the certified sites keep records of all customer complaints and how such complaints are managed and resolved.

In addition, the Standard advises us in 8.5.2: Corrective Action that we are to create a documented procedure to define how we manage customer complaints.

We have generally found that this set of directives is misunderstood. What is called for is a clear statement with regard to how customer complaints are managed and recorded. This could be done by means of corrective action reports, marketing and sales logs and memos, or any combination thereof. For those who work in the medical industry, this directive is simply part of the general FDA/CGMP requirement.

The prescriptive response to customer complaint management belongs as a separate section in the manual as part of Clause 8.5.2: Corrective Action. This process can then be referenced in Clauses 5.2: Customer Focus and 8.2.1: Customer Satisfaction as one of the methods to measure either customer satisfaction or dissatisfaction.

The partial redundancy of Clauses 5.2, 8.2.1, and 8.5.2 are an example of what we refer to as the Standard's concomitant relationships. The redundancy inherent in the Standard does cause some agony when you create the manual because you can easily end up repeating paragraphs. This can be ameliorated somewhat if you use references back to previous text to avoid this trap. Then flesh out the section with any really new requirements over the redundant requirements.

9.6.2 Registrar-Mandated Factored-Items Requirement

We can define a factored item as a product, shipped to your customer with your logo on it, that has not been manufactured under your certified quality management system. The existence of a factored item requires that the supplier alert the customer base to the fact that the product was not manufactured within a certified QMS.

An example of a factored item is a can of some chemical that you might stock and sell to your customers for their convenience. The can is purchased under private label from the manufacturer and then either inventoried in your shop and shipped from stock, or drop-shipped to the customer by the manufacturer. Although you do check the label for accuracy, you do not verify the product's specifications or integrity. The sale of such cans represents a significant percent of your total sales (e.g., 1% or more of total revenue).

As a result, this product must be declared as a factored item and cannot be included under those products that are processed through your ISO 9001:2000 quality management system. Your customers will need to be made aware of such a situation in some manner (e.g., a memorandum or a note in a catalog). Because brochures and catalogs are usually printed in the thousands, it is customary to use stick-on labels to correct the current documents until the next printing run. In such cases, it is best to inform your registrar of the issue and come to a decision on the best course to follow. The registrar will need to make the final decision [6].

If you do not wish to have factored items, the remedy could be just a simple sampling plan or could be as serious as a resident QC inspector at the manufacturer's plant. Other methods include: buying the material from an ISO 9000 certified and accredited supplier; periodic vendor audits; and certificates of compliance or analysis.

The declaration of a factored item(s), and a description of the process that is used to inform customers, is usually placed in Section 4.1: General Requirements.

9.7 Mandated Standards and Codes Requirement

The Standard is somewhat nebulous with regard to required standards and codes. There are several statements in ISO 9001:2000 Clauses 7.2.1 and 7.3.2 that, taken together, give some indication of the intent (i.e., they address statutory and regulatory requirements). What is advisable and appears to be generally acceptable to registrars is to create a list of the product-/process-oriented standards and codes maintained by the organization, state who is responsible for them, show where they are kept, and explain how they are kept current.

Table 9.2 summarizes various standards and codes that might be listed by an organization. The list will vary significantly based on the organization's product lines and certifications, which could include environmental, medical, and telecommunications, as well as automotive and aerospace requirements. For example, customer standards are numerous in the automotive industry. Standards and codes are normally very clearly defined in marketing requirements, engineering specifications, technical files, test procedures, and product brochures.

Table 9.2
Typical Standards and Codes To Be Listed

Name of Standard and Code	Responsible Manager	Where Located	Currency Method
21 CFR, Part 210	Regulatory affairs	Office files	U.S. government publications
ISO 9000:2000; Q9001-2000; & Q9004-2000	ISO 9000 management representative	Office files	Subscription to the ASQ
AAALAC guidelines	Pathology	Laboratory files	AAALAC committee member
NIH guidelines	Operations	On-site files	Subscription
U.S. pharmacopoeia national formulary	Quality-assurance manager	Laboratory files	Subscription to USP-NF
IPC-610 manufacturing codes	Quality-assurance manager	Design and manufacturing areas	IPC Membership
CE Mark: EN60601-1-2 EN55011 Class B IEC 801-2 IEC 801-3 IEC 801-4 IEC 801-5	Engineering-design administrator	Administrator's office files	Review annually and obtain latest revisions from test house
UL 2601	Design engineering	Design-engineering files	Subscription
Big Three automotive standards	Design engineering	Design-engineering library	Supplied by Big Three
Mechanical contractor standards	Senior engineer	Contractor's engineering library	Subscription

Endnotes

[1] I am not the only one concerned with this requirement. See, for example, Page, Stephen B., "The Continued Importance of Documented Procedures to ISO 9000:2000 Standards," *Software Quality,* Summer 2001, p. 13.

[2] The use of checklists has begun show up as articles. See, for example, Kaganov, Mark, "Checklists—A Perfect Tool To Tune Up Your Quality Manual," *Quality Progress,* October 2000, p. 37.

[3] See "Quality System Requirements, QS-9000," obtained from AIAG at tel: (810) 358-3003.

[4] See, for example, Taylor, C. Michael, "Bored with the Same Old Standards Books?" *Automotive Excellence,* Spring 1997, p. 6.

[5] As with all powerful ideas, an ISO 9000 mythology has been created in spite of its short 15-year existence. One of the myths is that each tier must have a document for each mandatory, and for that matter, each implied requirement. We feel that such an approach is not only contrary to the spirit of the Standard, but the redundancy that results from such a viewpoint is counterproductive and serves to confuse the users instead of support their efforts. The new revision attempts to remedy this attitude but may have oversimplified the requirements.

[6] Although the supplier is ultimately responsible for the choice of exclusion and how that exclusion is justified to their organization, it is essential to keep in close touch with your registrar on interpretation because the registrar has a similar issue (i.e., how the Standard should be interpreted against the requirements of the accreditation board).

PART

III

QMS Implementation

(George) Miller (1956) showed that the individual's ability to make absolute distinctions among stimuli, to distinguish phonemes from one another, to estimate numbers accurately, and to remember a number of discrete items all seemed to undergo a crucial change at about the level of seven items. Below that number, individuals could readily handle such tasks: above it, individuals were likely to fail. Nor did this discontinuity seem accidental.

—Howard Gardner, The Mind's New Science, New York: Basic Books, 1985, p. 89.

The beginning of wisdom is calling things by their right names.

—Old Chinese Proverb.

CHAPTER

10

Contents

The Quality Manual Scope of Effort

10.1 Estimates

A considerable effort is required by top management to produce a stand-alone ISO 9001:2000 sequenced manual that integrates business strategy with quality management. It is an iterative activity that peaks approximately one-third of the way into the process and then requires some level of maintenance up to the certification assessment. After certification, maintenance is normally required prior to a surveillance assessment and when the organizational and operational structure makes significant changes.

We can estimate to some degree the number of hours required to create a fully compliant manual if we assume there are the following:

- A process manufacturing facility;
- A staff of 100 employees—20% of which are managers and line supervisors;
- A quality-assurance department;
- A management representative who is also a full-time manager;
- A full-time clerical support;
- A part-time consultant (approximately 25% of the time on site during the precertification effort);

- A training program that includes documentation-writing skills for some employees;
- A documentation system that already exists in the form of some basic work instructions and operational formats;
- A plan that shows that the designated employees write, edit, and research for three hours for every hour that the consultant had been on site.

10.2 Discussion

The estimate scales with size and product complexity, so plus 50% and minus 20% is possible. Table 10.1 illustrates a typical scenario and plan for the manual. The time to certification assessment is 12 months from the program kick-off date.

As indicated in Table 10.1, to create a manual of approximately 50 pages requires a considerable effort of the entire staff—approximately 56 employee days. This is not a one-two-three exercise, and the effort includes team meetings and considerable dialogue. This estimate assumes that the development of processes has been completed before work begins on the manual.

As indicated, the load is greatest on quality assurance because we have assumed that at least internal quality audits and metrology have been assigned to that group, along with inspection and testing. The potential loading on each department will become clearer as we proceed through the rest of the text.

The result of such an effort is a manual that makes sense to all of its readers and propagates a favorable impression of the organization both from a strategic and technical standpoint.

Table 10.1
Excellent Corporation's Quality Policy Manual Timeline

Manual Phases	Scheduled Months for Actions in Gray											
Months from kick-off	1	2	3	4	5	6	7	8	9	10	11	12 Cert
Initial drafts due												
First draft review												
Final draft review												
First master published												
Master review after continuous improvement audit												
Master review after readiness assessment												
Master review after certification audit												
Total writer/editor/ research days												
ISO management representative	32	8	8	4		2		2		2	8	2
Technical writer		40	16	4		2		2		2	4	2
Clerical		40	16	8		4		4		2	8	2
ISO administration subtotals (hrs)	32	88	32	16		8		8		6	20	6
General manager	8	4	2	1		1		1		1	1	1
Engineering manager	12	6	3	2		1		1		1	1	1
Operations manager	12	8	4	2		1		1		1	1	1
Purchasing manager	8	4	2	1		1		1		1	1	1
QA manager	16	12	8	6		4		2		1	4	2
Marketing and sales manager	8	4	2	1		1		1		1	1	1
HR manager	8	4	2	1		1		1		1	1	1
Finance manager	4	2	1	1		1		1		1	1	1
Supervisors			16	4		1		1		1	1	1
GM and staff subtotals (hrs)	76	44	40	19		12		10		9	12	10
Grand total (hrs)	108	132	40	19		12		10		9	12	10

Grand total of hours = 448 employee hours
ISO administration = approximately 27 days
Approximately = 56 employee days
GM and staff = 232 hours = approximately 29 days

CHAPTER

11

Contents

Hub Documents

11.1 Definition

A handy universal bucket (hub) document is similar to an airport hub in that it is a center of information flow. The manual need only reference this one document in each major clause, and the hub document will then take the reader to the appropriate supplementary lower tier documents.

For example, in a structured, hypertext system (i.e., online), there would be only one icon per key clause of the Standard. Then, once you are in that referenced document, other icons would transfer you to the appropriate supplemental document. It is unnecessary to have a hypertext system to structure a hub system. This degree of simplicity is available regardless of the word processor design.

Section 11.2, which follows, offers an example of how a system can be simplified and made more user friendly without compromising the system's basic integrity.

11.2 Hub Template

A typical master reference list of a hub document–oriented manual is shown in Table 11.1. Each section of the manual would refer to the appropriate hub document. An alternative method of display would be, for example, by means of a documentation tree (see Figure 11.1).

Some key attributes of the hub design are as follows:

Table 11.1
Typical Hub Document References (Simplified)

ISO 9001:2000 Section	Manual Section	Typical Hub Documents	Typical Document Champions/Authors
4.0: Quality management system	4.1, 4.2.1	Business processes manual	Executive committee
		Security manual	Security officer
		Employee manual	HR manager
	4.2.2	Quality policy manual	CEO
			ISO management representative
			ISO management representative
	4.2.3	Control of documents procedure	
	4.2.4	Control of records procedure	
5.0: Management responsibility	5.1, 5.2, 5.3, 5.4, 5.5, 5.6	Business plan	Executive committee
		Quality policy directive	CEO
		Quality objectives directive	COO
		Organizational chart	COO
		Management review procedure	ISO management representative
6.0: Resource management	6.1	Resource management process	COO
	6.2	Human-resources manual	HR manager
	6.3, 6.4	Facilities manual	Plant manager
7.0: Product realization	7.1	Product realization process plan	Vice president operations
	7.2	Marketing and sales process manual	Plant manager
	7.3	Design and development standards	Vice president engineering
	7.4	Purchasing manual	Purchasing manager
	7.5	Operations manual	Vice president operations
		Customer service manual	Service manager
	7.6	Metrology manual	QA/RA manager
8.0: Measurement, analysis, and improvement	8.1, 8.2.1, 8.2.2, 8.2.3, 8.2.4	Monitoring and measurement control plan	Vice president operations
		Auditing manual	QA/RA manager
	8.3	Nonconforming product procedure	QA/RA manager
	8.4, 8.5.1	Statistical analysis manual	QA/RA manager
	8.5.2, 8.5.3	Corrective and preventive action procedures with customer complaints	QA/RA manager

Figure 11.1 Excellent's hub documentation linkage tree (simplified).

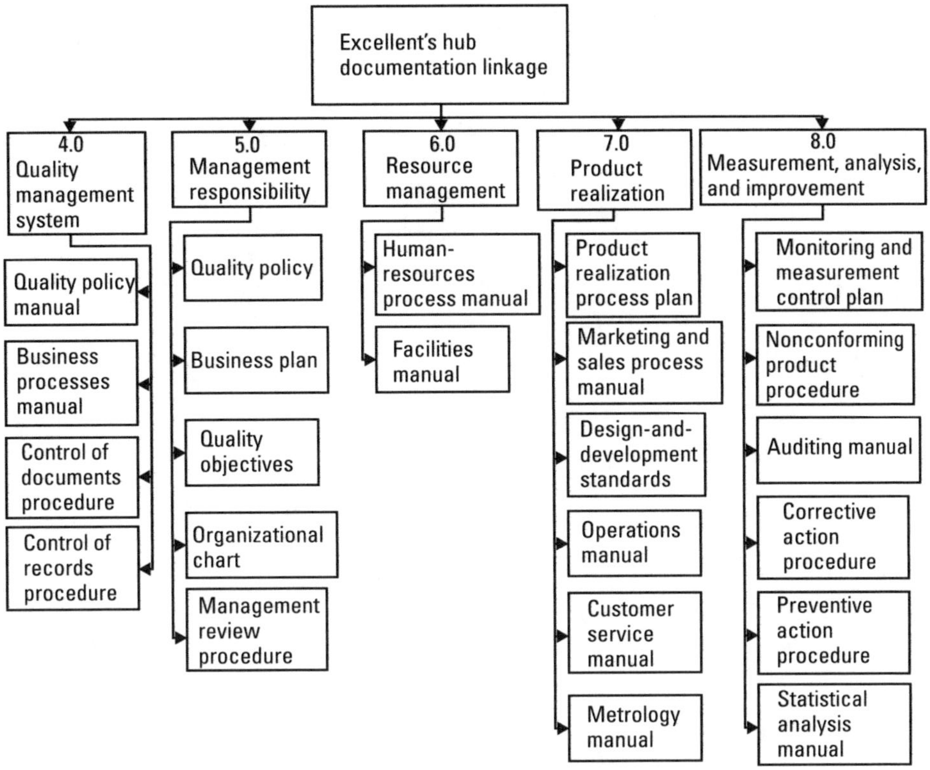

- In most cases, there is only one hub document per major clause (e.g., Clause 6.2: Human Resources Process Manual).
- Several hub documents are used within a section when they either represent unique processes or require a summary statement (e.g., the business plan is a partial response to all of Section 5.0 and is a hub document because of its scope and multiple topics).
- An operations manual is used in Section 7.0 as a hub document to contain the various procedural requirements for the production process.
- A service manual is used in Section 7.0 to cover after-sales processes.
- Hub documents are invariably very high level process or standard operating procedural documents, including quality and control plans.

CHAPTER

12

Contents

Quality Manual Issues

12.1 Hard-Copy Manual Issues

12.1.1 Manual Control

In practice, a pure hard-copy system is the most expensive and time consuming to maintain, and it is best to limit the number of controlled manuals to essential personnel [e.g., document owner (often the site manager), ISO 9000 management representative, and the registrar].

Uncontrolled copies usually need to be released by the owner on a filtered basis for marketing and educational purposes. However, the manual should have some sort of disclaimer (e.g., "The contents of this uncontrolled manual may not be at the latest revision level"). Because the manual is usually revised on the average of about twice a year (e.g., after a surveillance assessment and after organizational changes), the currency of the document is not a big issue and uncontrolled copies are not really a concern.

12.1.2 Manual Revisions

It is important to minimize the number of times per year that changes are made to the manual to minimize printing costs. Such costs can be very significant when you consider the cost of labor and distribution control. It is best to collect minor changes and do a rewrite periodically unless, as we pointed out earlier, there has been some major action taken (e.g., reorganization, third party audit that resulted in nonconformances, merger/acquisition activities, or a business scope upgrade).

12.1.3 Manual Distribution

The creation and distribution of the manual must comply with the requirements of Element 4.2.3: Control of Documents of the Standard. A convenient checklist to be used to ensure this compliance is shown in Appendix E, entitled "Checklist for Standard Element 4.2.3: Control of Documents Quality Manual Requirements."

12.2 Online Manual Issues

12.2.1 Impact of the Online Manual

With the advent of enhanced information technology networks, many organizations are either already networked or plan to be in the near future. Any move to place the manual online will have an immediate impact in the ease of control. The amount of software available for online use is overwhelming. The platforms are either self developed or based on readily available software [1]. It is not uncommon to find both the certificates of registration and the manual on an organization's website.

However, as the manual serves as an excellent marketing tool, we will still want to produce uncontrolled hard copies under the same conditions mentioned earlier. In other words, an online manual tends to always end up a mix of electronic and hard-copy media. This is often true for the entire documentation system because we find that drawings, blueprints, schematics, data sheets, and production tags, for example, tend to remain as hard copy, especially in smaller companies. Larger companies tend to favor more electronic files via scanned documents, but this requires an extensive and sophisticated computer system.

12.2.2 Key Factors

The decision to go online involves the solution of a number of critical factors, several of which are beyond the scope of this text. However, a few examples of some key factors include the following:

- *Structured hypertext [2]:* The use of hypertext alone will not guarantee an effective system unless the entire documentation structure is logically designed on the basis of hierarchal need. The old adage, *garbage in, garbage out,* still holds true. The online manual's cover page is an excellent location to place hyperlinks, not only to the manual's sections, but to the master lists for all the tiers.

- *Available expertise:* Even if the choice is made to go with off-the-shelf quality management system software (QMS/W), we have found it necessary to have someone on board who is a computer expert, in conjunction with a dynamic training program. Most importantly, there is a clearly defined need to have support available 24 hours a day, 7 days a week. The reason for this is that QMS S/W packages are designed to manipulate ideas as opposed to MRP- and SPC-type packages that are designed to manipulate data. As a result, there is a constant need for clarification as to what the information means when ideas are involved. Also, unless you have personally designed the QMS S/W databases you will be ill equipped to correct logical software glitches.

- *Graphics (flow charts, tables):* Although graphics, and in particular flow charts, can greatly enhance the overall effectiveness of a document's usefulness, unless there is clear evidence that the flow charts can be effectively integrated into the document's application software, it may be better to use tables as a means of clarity. It is always best to know the limits of interoperability for your software before you invest a great deal of time and funds into any type of graphics.

- *Training issues:* The moment the decision is made to go online, the training must begin immediately. As we noted previously, it has been our experience that online systems require far more training than hard-copy systems.

- *Projection systems:* To avoid an unacceptable level of dropped hard copy in an online system (e.g., for meetings or training sessions), it is advisable to install projection systems that are driven by your computers. The issue is one of projection intensity, and it needs to be checked out before installation to keep everyone in the room from dozing off in front of the president. Modern projection systems (somewhat costly) have intense light capabilities, so this problem should no longer exist.

- *Online impact on registrars and assessors:* Presently, online QMS documentation systems are quite common, but you will find a wide range of methods used by third-party assessors to accept and recommend certification for an online system, particularly when document usage is often performed at the customer's or a distant sales office's site. It is best to check in with your registrar and develop a mutually agreeable audit plan that will resolve this issue. I have been able to use downloaded manuals for this purpose from my clients without trouble. Just keep the number of documents transferred on the Web to a reasonable number.

Endnotes

[1] The ASQC publication *Quality Progress* issues a yearly summary of available software. See, for example, "2002 Software Showcase and Directory," January 2002. Also, "2002 Quality Sourcebook," *Quality Digest,* January 2002.

[2] An excellent discussion on structured hypertext is given in Horn, Robert E., *Mapping Hypertext,* Lexington Institute, Information Mapping, Inc., Waltham, MA, 1989.

CHAPTER

13

Contents

Leadership

13.1 ISO 9000 Stewardship

Clearly, the QMS needs an organizational home [1]. It requires the following:

- Ownership and oversight by top management;
- A way to be created, controlled, and revised;
- Acceptance by all users;
- Evidence that it is worthwhile.

Unfortunately, for most users, the QMS is something that suddenly appears and is overwhelming with its unaccustomed vocabulary and demands. As a result, it is important to create a documentation system that

- Is worth reading;
- Contains phraseology familiar to the industry;
- Is relatively easy to work with (user friendly);
- Truly represents the policies, processes, procedures, and formats of the organization.

For example, in the case of the manual, we have assumed that the manual is actually distributed in such a way that all QMS users can obtain a copy if they so desire. We have found that this is not always the case. We believe it should be, especially as the rest of the documentation system is made available to those users who need them on a daily basis. I find it unfathomable that a document that states the strategic position of the enterprise should be considered not appropriate for the average user to read. Interestingly enough, Par. 5.5.3: Internal communication of the Standard requires that top management ensures that appropriate communication processes be established within the enterprise to alert users on the effectiveness of the QMS. What could be a more effective way to establish a framework for such discussion than a readable, reality-oriented manual?

Our intent is to make the QMS fully compliant with the Standard's clauses so that we gain the benefit of the inherent interplay, whereby one clause either supports or relates to another clause.

In particular, we wish to make the manual fully compliant with the Standard so that it drives all of the other documentation levels in this direction. We want the manual to reflect the organization's dedication to an integrated business/quality theme.

We have found an effective way to ensure that the manual is compliant and distributed appropriately and represents the organization's personality. This is to assign various staff members (stewards) with specific sections of the Standard so that they are responsible for the documentation, implementation, and demonstration of effectiveness of each ISO 9000 clause down through all operating levels of the system. The stewards may take on roles such as process steward, subprocess steward, and section steward. This approach is even more important than it had been in the previous revision because the Standard now requires that quality objectives be established at relevant functions and levels within the organization (Par. 5.4.1: Quality Objectives).

As a result, it is not just that top management establishes quality objectives, but that operational areas also establish objectives that support the top-level objectives.

Multilevel Quality Objectives We will now establish the various roles and duties assignable to the stewards. The requirement for documentation, implementation, and demonstration of effectiveness—that forms the three pillars of ISO 9000—is illustrated in Figure 1.3 and is based directly on Clause 4.1: General Requirements of the Standard, which directs us to create and implement a documented QMS and to continually improve its effectiveness.

The interpretation of this requirement, to the effect that the prime directive of ISO 9000 is documentation, has been the most maligned in both the 1987 and 1994 versions and is already going in the wrong direction for the 2000 version. It is patently not true that the prime requirement in ISO 9000 is documentation. Documentation now is, and has always been, at most, a third of the mandatory directives.

The total implementation of the QMS, and the demonstration that what has been implemented is effective in enhancing enterprise performance, has always been the prime directive. The 2000 version makes this point abundantly clear via its clearly stated requirement to continually improve QMS effectiveness.

From an engineering standpoint, the gain of the system (output divided by input) can be described quantitatively to some degree prior to certification by measuring how close we are to completeness. A gain of near unity (i.e., when we reach 90% of our documentation, implementation, and demonstration of effectiveness goals) can be used to successfully determine when it is time for the initial assessment by a registrar.

At the 90% point, the QMS is fully operational and ready for fine tuning. However, the fine-tuning process is never ending, so it is best to follow the rule of diminishing returns (i.e., move on to other tasks when it takes twice as much effort to go half as far as you previously traveled). When you reach this asymptotic behavior, the QMS needs to be shaken periodically to see what falls out, and that is done through the internal audit process. That is what "taking our temperature" is all about. It is the essence of Section 8 in the Standard, which contains, for example, both the internal audit and corrective and preventive action.

After certification, the gain can actually exceed unity and is based on how well we achieve our quality objectives, which are measured according to metrics that fit our industry. For example, assume that customer satisfaction is our gain measure and the metric is the dollar value of the contract awarded (for each new award) divided by the previous award's dollar value. We could see a dramatic increase in gain if the customer has shifted the previous percentage of awards from 20% to us and 80% to our competitors to 40% to us and 60% to our competitors based on our improved performance. Each enterprise will of course have its own appropriate gain measures and metrics. However, this is a two-way street, and the gain could be less than unity if our competitor is more interested in improved performance than we are.

13.2 The Stewards Take Our Temperature

The key to a successful QMS—one that helps us to achieve our quality goals and objectives—is the ability to measure either our progress against those goals and objectives or the lack thereof! Someone has to be responsible for this task. A useful chart for this purpose is shown in Table 13.1.

Table 13.1
Taking Excellent's Temperature—ISO 9001:2000 Readiness Chart

C/I Elements	Relative Percentage (%)								
Month	**Mar.**	**June**	**July**	**Aug.**	**Sept.**	**Oct.**	**Nov.**	**Dec.**	
Goal	**Launch Phase**	**First Draft Process Documents**	**First Audit Verification and Validation**	**Final Draft QM**	**Management Review**	**Internal Readiness Assessment**	**Registrar's Document Review**	**Registrar's Certification**	**Activities Where Points Will Come From**
Plan—executive									
Management reviews	30	50	50	60					2 Management reviews (15)
Quality manual	5	70	80	85					Readiness audit
Objectives metrics	30	50	60	70					Quantitative analysis
Demonstration of effectiveness	5	50	55	60					Full corrective and preventive action with customer complaints
Do—operational									
Process documents	5	60	65	75					Complete marketing and sales, service
Procedural documents	30	40	50	55					Readiness audit
Format documents	30	50	60	70					Production, test
Implementation	30	40	50	55					Engineering, marketing and sales
Master records lists	5	30	50	70					Production, test, service
Master documents lists	5	30	50	60					Production, marketing and sales
Approved vendor lists	5	30	70	75					Purchasing, engineering
Master calibration lists	5	30	50	50					Internal calibrations
Training program	30	40	50	55					Production, marketing and sales
Check—internal									
Customer satisfaction	30	50	60	65					Survey analysis
Quality audits	5	30	50	60					5 points each
Verification process and product	30	50	80	85					Readiness audit

Table 13.1 (continued)

C/I Elements	Relative Percentage (%)								
Month	**Mar.**	**June**	**July**	**Aug.**	**Sept.**	**Oct.**	**Nov.**	**Dec.**	
Goal	**Launch Phase**	**First Draft Process Documen ts**	**First Audit Verifica-tion and Validation**	**Final Draft QM**	**Manage-ment Review**	**Internal Readiness Assessment**	**Registrar's Document Review**	**Registrar's Certification**	**Activities Where Points Will Come From**
Act–effectiveness									
Data analysis	5	35	50	70					Complete intranet
Corrective actions	5	40	50	65					Analyze
Preventive actions	5	40	60	60					Make a list
Customer complaints	30	50	60	65					Analyze
Continual improvement	30	50	60	65					Trending

Note: 90% implies readiness for the initial assessment.
Guide: startup required = 5%; a clear base exists = 30%; mid-game to ISO readiness = 35–70%; endgame, the last few yards = 75–85%; ready for initial assessment = 90%.

In Table 13.1, we use the Shewhart cycle of plan-do-check-act to create a quantitative matrix for this measurement purpose. The paradigm has been given a parallel terminology, (i.e., executive-operational-internal effectiveness to emphasize the role that is to be played by management).

In the example, the Excellent Corporation, which started from a fairly weak base, has progressed month by month to the point where the manual is about one more tight-end football pass away from a touchdown and some exceptional work has been done in the verification of product.

A key duty of the steward is to input the scorecard. The scorecard is kept up to date weekly by the ISO 9000 management representative based on inputs from the stewards. Thus, we have defined the first key duty of a steward.

To further determine the duties of stewardship, we note that in Clause 5.1 of the Standard, it is required that top management clearly indicates its commitment to QMS documentation, implementation, and effectiveness. A continually improved QMS effectiveness is to be demonstrated. The Standard then lists a number of directives that can be used to create a matrix that defines stewardship (see Table 13.2).

Thus, the essential role of a steward (process champion) is to ensure that the QMS contains the following:

- An effective and visible quality policy statement;
- A program management plan to guide the team's activities;

Table 13.2
Duties of the Steward

ISO 9001:2000 Clause	ISO 9001:2000 Requirement	Related Specific Duties of the Stewards with Their Action Teams
5.1(b)	Quality policy: establish the quality policy	Formulate the quality policy statement (QPS) Post the signed and controlled QPS Devise methods to propagate the QPS
5.4.2(a)	QMS planning: ensure that QMS planning meets the requirements of 4.1	Program manage the quality manual Program manage process documents Program manage procedures and forms Determine effectiveness guidelines Provide resources to complete QMS Determine, monitor, measure metrics Determine continual-improvement metrics
5.4.2(b)	Revision control: ensure that QMS integrity is maintained when changes are planned and implemented.	Ensure that all documentation is controlled
5.4.2(a)	Quality objectives: ensure that QMS planning meets the quality objectives	Plan in quality objective metrics into the QMS program management plan
5.1(c)	Quality objectives: establish the quality objectives	Formulate quality objectives at all levels that support top management goals
5.5.1	Responsibility and authority: define responsibilities and authorities and communicate them within the organization	Establish the sections of the Standard that each steward is responsible for and set up teams
5.1(e)	Resource management: ensure the availability of resources	Include all required resources into the program management plan
5.5.3	Communication: ensure that appropriate communication processes are established within the organization	Devise a communication process to inform the steward team members and top management on progress
5.1(a)	Communication: communicate to the organization the importance of meeting customer, statutory, and regulatory requirements	Include weekly reports that cover all aspects of the QMS creation, including the coverage of customer, statutory, and regulatory requirements
5.5.3	Communication: ensure that communication takes place regarding the effectiveness of the QMS	Include weekly steward team reviews in the program management plan
5.2	Customer satisfaction: ensure that customer requirements are determined and met with the aim of enhancing customer satisfaction	Include a metric for customer satisfaction in the program management plan
5.5.2	Management representative: appoint a management representative	Appoint an ISO 9000 management representative
5.1(d)	Management reviews: conduct management reviews	Prepare and present steward team status at each top management review

- A controlled documentation system;
- A set of quality objective metrics and goals at all levels of the organization that includes customer satisfaction metrics;
- A clearly defined set of team responsibilities and authority to document, implement, and demonstrate the effectiveness of the QMS via verification and validation analysis;
- Clear channels of communication to all groups involved in the creation process;
- Documented appointment of an ISO 9000 management representative;
- A dynamic presentation role in the top management reviews to clearly define the status of the QMS creation process for the steward's team.

Steward's Information Objectives With regard to communication, each steward ensures the following of their channels of information:

- There is a technical correlation from quality policy statement down to the lowest documentation levels (e.g. forms—the channel).
- There is an effective link between each quality policy statement and the lower tier documents.
- The entire documentation system is complete.
- The documented system is completely implemented.
- The channel is helping to achieve the quality objectives.
- The continual improvement programs are effective.
- There is an acceptance at all levels of the QMS directives.
- The quality policies are managed at all levels.

13.3 Team Leaders

It is also customary for the stewards to assign cross-functional teams to handle the more broad reaching requirements that touch several operational areas. The leader for such a group is sometimes called the team leader and for more complex activities can hold the title of program or project manager. Their responsibility is to ensure that specific cross-functional programs are effectively managed. Such programs include the following:

- *ISO 9000 management representative:* usually the quality-assurance manager, if such a position exists. Often found to be the president or head of operations when the QA function is not formally designated. Lower level employees are sometimes used with a dotted line authority to the top manager. (They are often responsible for 4.1: General Requirements and 4.2: Documentation Requirements using action teams as support groups. Responsibility includes the coordination of the total QMS creation process through the certification audit.)

- *CAPA manager:* usually the quality-assurance manager if the function exists, and if it does not it is usually shared by several managers (e.g., the customer service manager stewards the customer complaints activities). Refer to Clauses 5.2: Customer Focus and 8.5: Improvement of the Standard.

- *Total audits manager:* usually led by the QA manager if the function exists, otherwise stewarded by another manager (e.g., controller, head of operations, director of safety). Refer to Clause 8.2.2: Internal Audit. This clause has again been written myopically. For a far more productive approach to the audit, it should include first-, second-, and third-party audits, not just first-party audits. For example, one of the most useful audits is your customer's second-party audit of either you or your subcontractors.

- *Training manager:* usually performed by the human-resources manager if that function exists, and, if not, it can be shared by all local area managers (e.g., all department heads are responsible for the training and documentation of their staffs). Refer to Clause 6.2: Human Resources.

These assignments can take many forms, but a possible distribution of responsibility in a design and manufacturing facility is demonstrated in Table 13.3. An effort has been made in the table to evenly distribute the stewardship responsibilities among top management. Unfortunately, it often happens that operations and quality assurance end up with an inordinate level of activity compared to the other departments. This type of situation is to be avoided when possible, as everyone in the company is usually already overloaded. Appendix A is a convenient way to catalog both the stewardship and team leader/project assignments.

13.3.1 Cross-Functional Team Organization

Table 13.3 demonstrated how the sections and clauses of the Standard can be generated by means of stewards and cross-functional teams. Although team establishment and organization is not an obvious dynamic, there is nothing

Table 13.3
Possible Stewardship Distribution by ISO 9001:2000 Section and Element in a Design and Manufacturing Facility

Responsible Steward	Responsible for These Sections	Action Teams Required	Team Leaders for Action Teams
General manager	4.1: General Requirements	4.1: Executive process team	General manager
	5.0: Management Responsibility	5.0: QMS planning and review team	
	6.1: Provision of Resources	6.1: QMS planning and review team	
	8.2.3: Monitoring and Measurement (M&M) of Processes	8.2.3: Executive process team	
	8.4: Analysis of Data	8.4: Team up with QA	
	8.5.1: Continual Improvement	8.5.1: QMS planning and review team	
ISO management representative [2]	4.0: Quality Management System	4.1: Process integrator document team	4.1 ISO management representative
		4.2.2: Quality manual integrator team	4.2.2 Document control specialist
		4.2.1: QMS integrator team	4.2.1 Document control specialist
		4.2.3: Document control team	4.2.3 Document control specialist
		4.2.4: Control of records team	4.2.4 Finance manager
Sales and marketing (S&M) manager	4.1: General Requirements S&M	Sales and marketing process team	Sales and marketing manager
		4.1: Team up with GM	
	7.2: Customer-Related Processes		
	8.2.1: Customer Satisfaction		
	8.2.3: M&M of Processes		
	8.4: Analysis of Data	8.4: Team up with QA	
Engineering manager	4.1: General Requirements Engineering	Engineering process team 4.1: Team up with GM	Engineering manager
	7.3: Design and Development		
	8.1: General		
	8.2.3: M&M of Processes		
	8.4: Analysis of Data	8.4: Team up with QA	
	8.4: Analysis of Data	8.4: Team up with QA	

Table 13.3 (continued)

Responsible Steward	Responsible for These Sections	Action Teams Required	Team Leaders for Action Teams
Purchasing manager	4.1: General Requirements Purchasing	Purchasing process team 4.1: Team up with GM	Purchasing manager
	7.4: Purchasing		
	8.2.3: M&M of Processes		
	8.4: Analysis of Data	8.4: Team up with QA	
Operations manager	4.1: General Requirements OPS	Operations process team 4.1: Team up with GM	Operations manager
	6.3: Infrastructure		
	6.4: Work Environment		
	7.1: Planning of Product Realization		
	7.5: Production and Service Provision		
	8.2.3: M&M of Processes		
	8.2.4: M&M of Product		
	8.4 :Analysis of Data	8.4: Team up with QA	
Quality-assurance manager	4.1: General Requirements QA	QA process team 4.1: Team up with GM	QA manager
	7.6: Control of Monitoring and Measuring Devices		
	8.2.2: Internal Audit		
	8.2.3: M&M of Processes	8.2.3, 8.2.4, and 8.3: Team up with operations process team	
	8.2.4: M&M of Product		
	8.3: Control of NC Product		
	8.4: Analysis of Data	8.4: CAPA team	
	8.5.2: Corrective Action	8.5.2: CAPA team	
	8.5.3: Preventive Action	8.5.3: CAPA team	
Human-resources manager	4.1: General Requirements Human Resources	HR process team 4.1: Team up with GM	Human-resources manager
	6.2: Human Resources		
	8.2.3: M&M of Processes		
	8.4: Analysis of Data	8.4: Team up with QA	
Finance manager	4.1: General Requirements Finance	Finance process map team 4.1: Team up with GM	Finance manager
	4.2.4: Control of Records	4.2.4: Team up with control of records team	
	8.2.3: M&M of Processes		

Table 13.3 (continued)

Responsible Steward	Responsible for These Sections	Action Teams Required	Team Leaders for Action Teams
MIS manager	4.1: General Requirements MIS	IT process team 4.1: Team up with GM	MIS manager
	8.2.3: M&M of Processes		
	8.4: Analysis of Data	8.4: Team up with QA	
Customer-service manager	4.1: General Requirements Service	4.1: Customer service process team 4.1: Team up with GM	Customer-service manager
	7.5: Production and Service Provision	7.5, 8.2.3, 8.2.4, and 8.3: Team up with operations process team	
	8.2.3: M&M of Processes		
	8.2.4: M&M of Product		
	8.3: Control of Nonconformance Product		
	8.4: Analysis of Data	8.4 Team up with QA	

Legend: M&M: monitering and measurement; CAPA: corrective and preventive action with customer complaints.

mystical about this approach, and each organization should structure their teams in a way that best fits their purpose. We have found it helpful when design and manufacturing teams form as shown in Table 13.4. The teams dealt with their appropriate clauses and SHALLS within the Standard's sections indicated. For additional detail, refer to Appendix A. Have some fun with the team names. They run all over the place from the Charlie's in honor of their leader, to Tool Time, for obvious reasons, and Enthusiasts for possible political reasons, and Road Runners to resolve poor vendor performance—and the reports even have baby pictures of the members. Team organization is replete with fanciful names such as the use of executive steering committees and area champions. Impressive action team awards and plaques are used to create the ownership required in response to the tremendous efforts put out by the teams. The results in bottom-line dollars saved are quite remarkable and over 5 to 10 years can be in the billions of dollars for multibillion dollar companies [3].

13.3.2 Organizations Without Explicit Design or Quality-Assurance Functions

For those organizations without an explicit design or quality-assurance department, both the Giants and the Seers Teams would be led by the manufacturing manager. The Giants would focus on process and product engineering themes that, although they do not include design, still require engineering discipline.

Table 13.4
Team Member Grouping

Action Team Members	Standard's Requirements
Tigers Site manager (leader) Sales and marketing manager Service manager	5.1, 5.2, 5.3, 5.4, 5.5, 5.6, 6.1, 6.2.1, 7.2, 7.5.1, and 8.5.1
Titans ISO management representative (leader) Document control administrator Finance manager	4.1, 4.2.1, 4.2.2, 4.2.3, 4.2.4,5.4.2, and 7.1
Giants Engineering manager (leader) Program managers Manufacturing engineers MIS manager	7.3, 8.1, 8.2.3, 8.2.4, and 8.4
Perks Purchasing manager (leader) Production control supervisor Engineering manager	7.4 and 7.5.4
Bulls Manufacturing manager (leader) Shipping/receiving supervisor MIS manager	6.3, 6.4, 7.5.1, 7.5.2, 7.5.3, and 7.5.5
Panthers Quality-assurance manager (leader) Manufacturing manager Facilities manager	7.1, 7.4.3, 7.5.3, 7.6, 8.1, and 8.2.4
Seers Quality-assurance manager (leader) Finance manager Manufacturing engineers Site manager Sales and marketing manager	8.2.1, 8.3, 8.4, and 8.5
Diggers Audit team manager (leader) Auditors Quality-assurance manager	8.2.2, & 8.2.3
Green Sox Human-resources manager (leader) Department managers	6.2.2

In this case, the Standard requires that we discuss clearly in the manual why Section 7.3: Design and Development is not required. Our prescriptive response would be something like this:

7.3 Design and development exclusion: The Excellent Corporation does not design and develop its products but receives this information in the form of a manufacturing release package from its customers. The manufacturing release includes the bill of materials, printed wiring board layouts, engineering change orders, and test procedures. Engineering support services are then provided by Excellent in the form of manufacturing engineering that includes both process and product engineers under the direction of the vice president of operations. Excellent is required to obtain authorization from its customers for any changes that affect form, fit, function, reliability, safety, or other regulatory or statutory requirements. The process related to manufacturing engineering is contained in the document entitled "Manufacturing Engineering Procedures."

13.3.3 Team Effectiveness

Figure 13.1 illustrates the need for cross-functional teams in the effective implementation of the Standard so that the quality improvement team (QIT) can effectively integrate the functional areas of corrective action, preventive action, customer complaints, and nonconforming product. For example, the QIT could consist of members from QA, engineering, manufacturing, customer service, and finance. The several interactive functions required for each key activity is also listed in Figure 13.1.

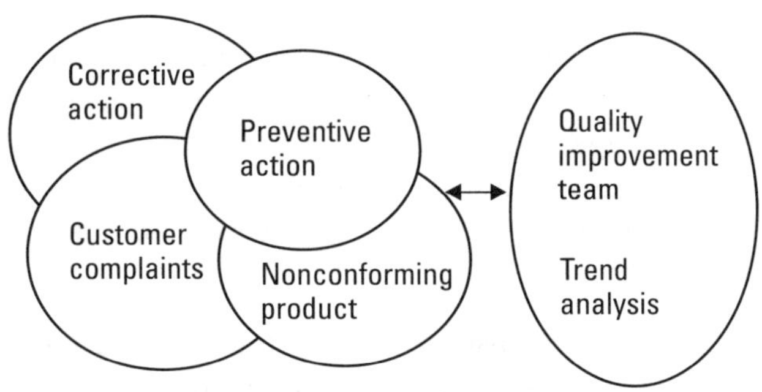

Corrective action
- Reactive
- Short range
- Band-aid
- Partial/interim
- CARs/SCARs
- Applies to:
  - IQAs
  - vendors
  - customers
  - registrars

Preventive action
- Proactive
- Long-range
- Root-cause
- Must report to executive
- Applies to:
  - extensive CARs
  - new designs
  - product improvements
  - organization
  - requires trend analysis

Customer complaints
- Must discuss
- Effective handling
- Keep records

NC product
- Product NCs
- Process NCs
- Nontraditional aspects

Figure 13.1 Corrective and preventive action with customer complaints and interface to 8.3: Control of Nonconforming Product.

We have found that this area requires the most intensive training and takes the longest to optimize. When one looks in detail at the requirements of Elements 8.3, 8.5.2, and 8.5.3, the difficulty in interpretation is not such a surprise.

We offer some suggestions in this regard:

- Use SCARs to manage the interface with subsuppliers (subcontractors) as defined at receiving/receiving inspection. Nonconforming material reports (NCMRs) are also often used for this purpose and limited to incoming material issues.
- Limit 8.3: Control of Nonconforming Product to nonconformance reports (NCRs) that occur after incoming (receiving) and prior to shipment.
- Limit corrective action reports (CARs) to the internal quality audit findings and for big-time nonconformances that require a team. By their nature, CARs are expensive and time consuming. Allow local area managers and supervisors to correct small-time issues with on-the-spot corrective actions. Keep a log of such actions for trend analysis.
- Assign one person to decide on the level of CAR responses (e.g., manager of QA). Filter out those that can be handled quickly without a lot of paper work and those that really need some time and effort and are worthy of documentation and trend analysis.
- Run the preventive action program via memos and reports. Stay away from a specific format—preventive actions by their nature are broad in scope and need a lot of creativity to carry them through to completion. Preventive actions should include significant improvements in organizational, material, facility, root-cause analysis, instrumentation, and MIS changes.
- Manage customer complaints via the returned material authorization, returned goods authorization, or returned authorization transactions, on logs kept by sales and marketing and maintained by customer service or an equivalent organization.
- Assign one person to filter the customer complaints for the required response (e.g., manager of sales and marketing). For regulatory organizations, the same person can make the decision to proceed if, for example, a medical device report is necessary. The complaint would then be immediately forwarded to the manager of regulatory affairs for investigation and action.

- Assign one person to take primary responsibility for the overall review and trend analysis of the databases (e.g., manager of quality assurance, manager of business development, manager of business operations). That person manages the preparation and reporting functions of the quality improvement team with regard to trend data and analysis.
- This function directly supports the requirement 8.4: Analysis of Data and 8.5.1: Continual Improvement.

13.3.4 Typical Real-Time Action Team Plan

Figure 13.2 illustrates a typical real-time action team plan that has been used successfully to achieve impressive action team performance. In this scenario, a facilitator is assigned by top management, and a team leader, scribe, and scorekeeper are also chosen. The scribe keeps the minutes and the scorekeeper calculates potential and real-time cost savings as the program

Figure 13.2 The continuous improvement cycle based on the Deming PDSA cycle (Deming 1986).

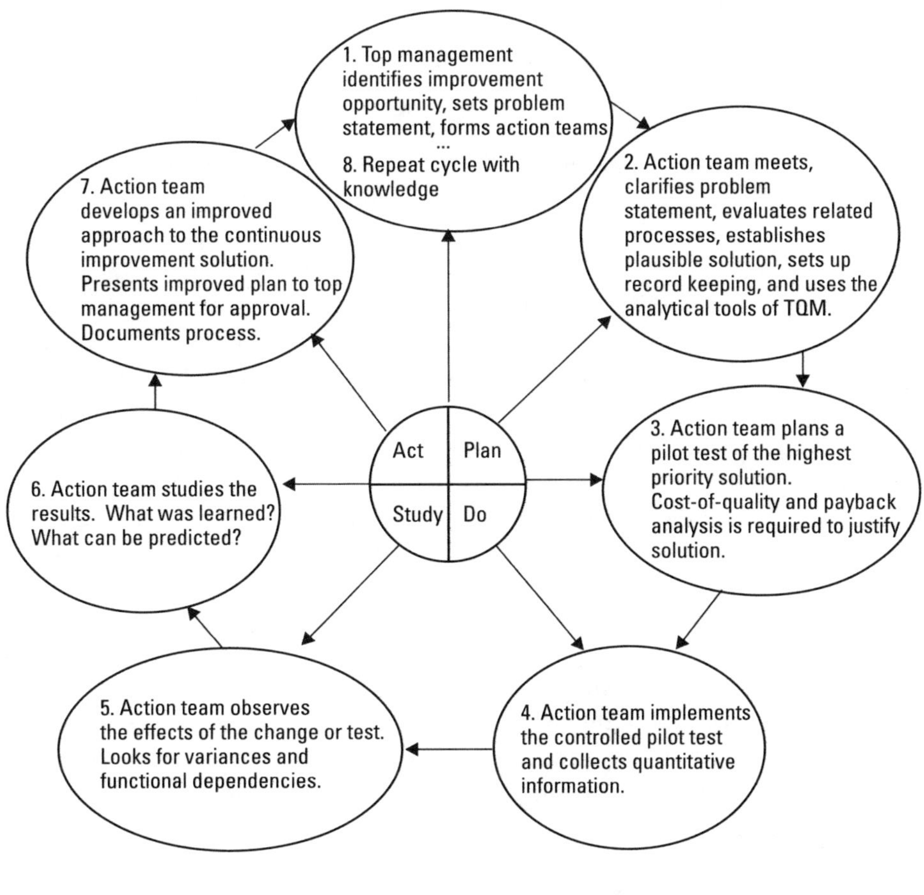

proceeds. If possible, the scorekeeper should be a member of the finance department to ensure that bottom-line considerations are prominent in the calculations [4].

All of these positions require extensive training to ensure a meaningful and productive team exercise. Top management oversight is required periodically to make sure that the team is coupled to the organization's objectives both technically and financially [5]. A team plan should use the following:

- The stewardship concept to effectively control QMS design;
- Cross-functional teams to ensure the accuracy and usefulness of documentation;
- Top management oversight periodically to ensure compliance with the enterprise's objectives when cross-functional teams are used.

13.4 Certification Audits

13.4.1 You Cannot Fail

> It is impossible to fail certification (unless you quit). The worst thing that can happen is that it might take a little longer and cost a little more.

The final point that we wish to make in our discussion of the direct sequence manual is that you cannot fail an initial assessment, unless you simply quit. The worst thing that can happen is that is might take longer and cost more. This is an established fact for the initial systems assessment (certification assessment). One does not fail a third-party assessment; it is a part of the ISO mythology. One does get nonconformances that need to be corrected. The worst case is a major finding that could delay the certification process by up to three months and cost some more to pay the registrar's lead assessor to come back and clear the nonconformance. But that is it. This is the primary reason that so many consulting groups will agree to guarantee certification/registration [6].

The steward's task is to make sure that there are no major findings possible. This is accomplished via in-depth internal audits by well-trained auditors. The audits should be evenly distributed throughout the creation process and not left to the last moment prior to the document review. The audits not only increase the probability of a major nonconformance-free certification assessment, but they form the base of a dynamic corrective and preventive action program.

Inevitably there will be minor findings at the initial systems assessment, the first surveillance, the second surveillance, the recertification assessment, and the re-recertification assessment. That is what continuous improvement is all about. I still come up with nonconformances with clients that I have audited for over 8 years.

Organizations undergo all manner of change over 3 years (e.g., top management changes; mergers; acquisitions; moves to new facilities; market ups and downs; national and international tragedies, including war, floods, and fires). Without sufficient audits, the documentation falls behind reality and even the act of auditing begins to evaporate. It is equivalent to firing the sales staff because sales are down. Find the root causes, make the necessary changes to match the changed scenario, and move forward.

There, of course, can be major findings. By major findings we mean, for example, an ineffectual management review, a poorly managed training program, a lack of internal quality audits, a corrective and preventive action program that is uncertain and loosely managed. The stewards must pay close attention to these areas. One of the traps in the management review process is for the top manager to use the management review as a "rah rah" session instead of focusing on the enterprise's deviations from its planned goals based on firm and quantitative metrics. You say, "Never happens"? It does.

Another danger area is the loss of internal auditors due to downsizing, burnout, disinterest, and promotion. It is important to maintain a constantly trained group of auditors to cover such contingencies. A safe level of auditors depends on the organization's size in both people and square footage and the degree of outsourcing. Today, we have situations where the organization consists of one person in the site and everything else is outsourced. Your registrar will work with you to cover this event. It does happen and people get certified.

13.4.2 Audit Focus

An experienced assessor pays special attention to the requirements in the following:

- *Section 4: Quality Management System*—In this set lies the superstructure of the QMS and where change is controlled, especially with regard to processes and continual improvement.
- *Section 5.4: Planning*—This determines how closely quality objectives are planned and measured.
- *Section 5.6: Management Review*—This somewhat prescriptive set of paragraphs contains the review of continual improvement drivers of internal

audits, customer feedback, process performance, product conformity, preventive and corrective actions taken, and the manner in which top management responds to required change and opportunities for improvement.

- *Section 7.3: Design and Development*—Special attention is to be directed to the design review, verification, and validation functions.
- *Paragraph 8.2.2: Internal Audit*—This looks especially at whether all areas of the organization have been audited against all appropriate paragraphs and the audits have included all pertinent regulatory requirements.
- *Paragraph 8.5.2: Corrective Action*—This applies especially the management of customer complaints.
- *Paragraph 8.5.3: Preventive Action*—This requirement indicates clearly the degree to which the organization is either reactive to nonconformances (e.g., performs root-cause analysis on a set of nonconformances reported during corrective action) or takes a proactive perspective (e.g., performs risk analysis and designs in safety and introduces best practices to all operating groups based on improvements in one group to prevent nonconformities [7]) not only during the initial assessment but at every subsequent surveillance assessment. It is customary for registrars to require management review, design and development, internal audits, review of customer complaints, and review of QMS document changes to be mandatory for some percentage of the surveillance audits (e.g., every 6 months for internal audits and every 12 months for the design and development).

Special attention to these requirements ensures that the continuous improvement cycle is maintained throughout the life of the ISO 9000 program. When the Shewhart cycle is enforced, the odds are very high that the supplier will derive the benefits inherent from an effective QMS [8].

13.4.3 Assessor Role

Indeed, the role of the assessor is to teach and clarify. If this goal is met, the assessor feels fulfilled at the end of a long and intense audit, and the client feels that the effort was worth it. Alternately, if the assessor feels that the goal is to catch the client, both parties will end up with a feeling of uselessness, and the client will begin to seek out other registrars [9]. That the audit findings must be substantive, and of value to the client, is the foundation upon which the ISO third-party schema will either continue to expand or eventually decline.

In the search for added value, my most effective rule is to ask the gut-oriented question: does the method sound stupid? If it sounds stupid, it is—try another approach. This works every time. I always consider whether my finding will be of economic value to the enterprise. There is a fine line between conformance to the Standard and worth to the client. No system is perfect to start with, and no system becomes perfect in the process. Organizations are in constant change through new products, new technologies, acquisitions, mergers, the vagaries of markets, and the potential horrors of nationalistic power mania.

It is vital that the organization continually stretch its processes for improvement but not stretch beyond its economic boundaries. The auditor can play an important role in this scenario. It is best to try to get inside the mind of the top executive and see what makes sense within the strategic parameters of the operation. Auditors with this perspective will find themselves welcomed back more times than not.

13.4.4 Structure of the Audit

To carry out an effective audit of the Standard requires that we apply the pertinent clauses of the Standard against every enterprise process. This also means that we also ensure that each subprocess is covered in detail. Table 13.5 uses the same core competencies as shown in Figure 1.2.

Our example, shown in Table 13.5, is based on a small organization hierarchy. We have assumed that the departmental processes contain the following subprocesses:

1. *Executive:* business plan, management review, and steering committee;
2. *Marketing and sales:* servicing, product managers, marketing, sales, and distributors;
3. *RDT&E:* research and development, design, product support, engineering change, and document and engineering records control.
4. *Operations:* QA&RA, manufacturing, production control, purchasing, inventory control, and shipping and receiving;
5. *QA&RA:* ISO management representative, document and record control, metrology, corrective and preventive action, audits, quality control inspection, reliability, and data analysis and trending;
6. *Finance:* human resources, management information systems, financial control and analysis, and cost of quality support;

Table 13.5
Audit Plan for a Typical Manufacturing Enterprise

Processes	1. Executive	2. Marketing and Sales	3. RDTE	4. Operations	5. QARA	6. Finance	7. Human Resources	8. Service
ISO Clauses								
4.0: Quality Management System								
4.1:					*			
4.2.1:					*			
4.2.2:					*			
4.2.3:			*		*			
4.2.4:			*		*			
5.0: Management Responsibility								
5.1	*							
5.2:	*							
5.3:	*							
5.4.1:	*							
5.4.2:	*							
5.5.1:	*							
5.5.2:	*							
5.5.3:	*							
5.6:	*							
6.0: Resource Management								
6.1:	*							
6.2:							*	*
6.3:				*				*
6.4:				*				*
7.0: Product Realization					*			
7.1:				*	*			
7.2.1:		*	*					*
7.2.2:		*						*
7.2.3:		*			*			*
7.3:			*					
7.4:				*	*			*
7.5.1:				*				*
7.5.2:				*	*			*
7.5.3:				*	*			*
7.5.4:		*		*	*			*
7.5.5:				*	*			*
7.6:			*		*			*
8.0: Measurement, Analysis, and Improvement								
8.1:		*		*	*			*
8.2.1:	*	*			*			*
8.2.2:					*			
8.2.3:	*	*	*	*	*	*	*	*
8.2.4:				*	*			*
8.3:				*	*			*
8.4:	*				*	*		
8.5.1:	*				*	*		
8.5.2:	*				*			*
8.5.3:	*				*			*

7. *Human resources:* hiring, training, and employee development;

8. *Servicing:* customer service, repair, and installation.

The chart suggests which clauses to apply to which process and thereby suggests which employees are to be interviewed. The planned date of the audit and auditors could also be placed in the box instead the star. Other usual audit activities are also implied, such as auditing the distribution of documents throughout the facility, auditing records in various file cabinets, asking employees what they believe the quality policy means and who they think is the ISO 9000 management representative, and examining the status of training.

Unfortunately, there is no end of concern with regard to the manner in which we are to audit either (1) the requirement that no procedure is required for many clauses, or (2) the sometimes extremely descriptive language of some clauses (e.g., Clause 7.5.5: Preservation of Product). This clause is about as short and sweet as you can get with regard to a most complex and extensive issue that includes electrostatic discharge protection, shelf-life control, and a number of different types of preservation coatings as well as packaging and delivery. Fortunately, the topic of audit management has received wide recognition and many authors offer sensible ideas on how to approach the subject [10].

To formulate such an audit structure, it is important to realize that this process-oriented scenario has an intrinsic hierarchal structure of the type shown in Table 13.6.

13.4.5 Audit Plan for Sector-Specific Requirements

We can demonstrate the impact of a sector-specific requirement on the certification audit by means of the audit plan for sections of 4.0: Quality Management System and 5.0: Management Responsibility, as illustrated in Table 13.7. Notice that the although the assessor seeks answers to additional questions above and beyond the basic issues in ISO 9001, the questions are quite similar. The additional topics are highlighted in italics.

Table 13.6
Possible Hierarchial Organizational Structures

Small Organization	Large Organization
I Total process	I Total process
II Departmental processes	II Divisional processes
III Functional processes (subprocesses)	III Departmental processes
	IV Functional processes (subprocesses)

Table 13.7
Sector-Specific Impact on ISO 9001 Audits—Example 1

ISO 9001:2000 Element	Base ISO 9001 Assessment	Sector-Specific QS-9001 Assessment	Sector-Specific CGMP 820 Assessment	Sector-Specific ISO 9000-3 S/W Assessment
5.0: Management Responsibility	9:30 Scope Management commitment] Customer focus Quality policy Quality objectives QMS planning Responsibility, authority, and communication Management representative Management review	9:30 Scope Management commitment Customer focus Quality policy Quality objectives QMS planning Responsibility, authority, and communication Management representative Management review *Business plan* *Analysis and use of company-level data* Customer satisfaction	9:30 Scope *Medical class* Quality objectives *Customer complaints* Management commitment Customer focus Quality policy QMS planning Responsibility, authority, and communication Management representative Management review	9:30 Scope Management commitment Customer focus Quality policy Quality objectives QMS planning Responsibility, authority, and communication Management representative Management review *Customer's management responsibility* *Organization and customer joint reviews*
4.0: Quality Management System	10:30 General requirements Documentation requirements Quality manual Control of documents Factored items Interface issues Currency of Standards and codes/statutory/ regulatory	10:45 General requirements Documentation requirements Quality manual Control of documents Control of records *Control plans* *Special characteristics*	10:45 General requirements Documentation requirements Quality manual Control of documents *Quality plans* Control of records with: *design history file (DHF)*	10:45 General requirements Documentation requirements Quality manual Control of Documents Control of records *Life-cycle planning* Factored items

Table 13.7 (continued)

ISO 9001:2000 Element	Base ISO 9001 Assessment	Sector-Specific QS-9001 Assessment	Sector-Specific CGMP 820 Assessment	Sector-Specific ISO 9000-3 S/W Assessment
		Use of cross-functional teams Feasibility reviews *FMEAs* Factored items Interface issues Currency of Standards and codes/statutory/regulatory	*Device master record (DMR)* *Quality system records (QSR)* Factored items Interface issues Currency of Standards and codes/statutory/regulatory	Interface issues Currency of Standards and codes/statutory/regulatory

As indicated, more time is needed in the sector-specific cases because there are more SHALLS to cover and there is an increase in concomitance (e.g., there are additional sections in QS-9000 compared to the five in the Standard [11]).

The manner in which the organization provides answers to the additional questions is in exactly the same way that quality policy statements are used to respond to each SHALL of the Standard. In a previous book we demonstrated this technique and took an example from each of the three specific sectors shown in Table 13.7 [12]. We have repeated this work because the technique is invariant under the many changes that standards are scheduled to undergo. As a result, the exact language of the quoted standard may change but the method remains valid.

This discussion includes a more recent set of requirements in the medical device industry (i.e., we will examine the specific impact of the FDA CGMP 820, EN46001:1996, and ISO 13485:1996 on a manual:2000). Table 13.8 illustrates how this second set of medical device requirements are intertwined for two typical ISO 9001:2000 sections. Note that at the time of this writing, both EN46001 and ISO 13485 were still in the ISO 9001:1994 format. This situation has already caused some confusion in manual:2000 creation. However, as we have seen, cross-reference charts provide a quick way to harmonize the requirements and do not invalidate the suggested techniques [13].

For completeness, the sector-specific requirements for software are also shown in Table 13.7 based on ISO 9000-3, the guidelines for the application of ISO 9001 to the development, supply, and maintenance of software [14].

Table 13.8
Sector-Specific Impact on ISO 9001 Audits—Example 2

ISO 9001:2000 Element	Base ISO 9001 Assessment	Sector-Specific EN46001 Assessment	Sector-Specific CGMP 820 Assessment	Sector-Specific ISO 13485 Assessment
4.0: Quality Management System	10:30 General requirements Documentation requirements Quality manual Control of documents Factored items Interface issues Currency of Standards and codes/statutory/regulatory	10:45 General requirements *Regulatory requirements according to class* Documentation requirements *Technical files* Quality manual Control of documents Control of records Factored items Interface issues Currency of Standards and Codes/statutory	10:45 General requirements Documentation requirements Quality manual Control of documents *Control of records with:* *DHF* *DMR* *DHR* *QSR* Factored items Interface issues Currency of Standards and codes/statutory/regulatory	10:30 General requirements *Regulatory requirements according to class* Documentation requirements *Technical files* Quality manual Control of documents Control of records Factored items Interface issues Currency of Standards and codes/statutory/regulatory
7.3: Design and Development +7.2.1(c): Statutory and Regulatory Requirements (S&R)	1:00 Design and development Planning Inputs with S&R Outputs Review Verification Validation Change control	1:00 Design and development *Procedure* Planning *Inputs with S&R and with safety* Outputs Review *Verification with clinical investigation* Validation Change control	1:00 Design and development *Procedure by class* *Planning with review and approval* *Inputs with S&R and with intended use* *Outputs with signatures* *Review with DHF* *Verification with DHF* *Validation with pilot runs, S/W risk and validation, production specification* Change control with DHF	1:00 Design and development *Procedure* *Risk analysis* Planning Inputs with S&R Outputs Review Verification Validation with clinical invest-igation Change control

13.4.6 Tip of the Iceberg

When the day of the initial assessment arrives, it is important to realize that the assessors' observations represent the tip of the iceberg (see Figure 13.3). They only see what they need to see in order to assure themselves that the supplier has a workable QMS that will most likely produce a reasonable payback in a reasonable time. At least 90% of the nonconformances lie below the surface.

You, of course, know exactly what they are, and the assessors rely on you to make those corrections as part of an effective QMS program—especially by means of the internal audit process and, indeed, where applicable, audits of your suppliers.

It is not uncommon to feel that you have fooled the assessors once they leave. On the contrary, if you have, it is really a case of biting your nose to spite your face. They saw it, but did not have the time to investigate. On the other hand, you know it is there. So you need to fix it.

Otherwise, you can bet it will be found in a surveillance audit. Worst yet, it is a hole in the system through which profit dollars fall—and that is the whole point of an effective QMS—to fill those holes.

13.4.7 Dynamics of the Initial Assessment

At the close of the initial assessment, the lead assessor recommends certification, either with or without condition. The registrar's executive board approves and issues the registration numbers and certificates. The several

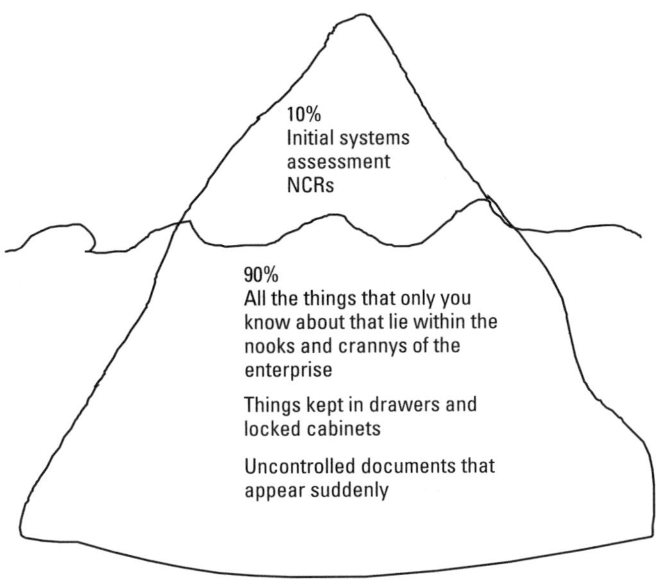

Figure 13.3 The tip-of-the-iceberg effect.

possible conditions for approval include the following (these vary considerably from registrar to registrar):

- All NCRs cleared during initial assessment—recommend certification without condition;
- Minors left to be cleared after initial assessment, but plans accepted—recommend, certification but hold issuance until all are cleared or hold clearance for first surveillance;
- Make sure there is a clear plan to be followed up at first surveillance;
- Some minors can be declared concerns to be monitored at the first surveillance;
- Opportunities for improvement—potential economic savings; these are to be acted upon at the discretion of the auditee.

The exception is in regard to major nonconformances. They are usually treated as follows:

- Majors left to be cleared during initial assessment require a return audit of those areas within usually 90 days, then recommendation to certify [15].
- Majors can be downgraded during the initial assessment to avoid this problem. The resulting minor can then be treated as discussed in the recommended-for-approval protocols. Downgrades are highly discretionary on the part of the lead assessor and must be examined in the context of the observed overall effectiveness of the audited QMS. Some registrars have strict protocols for downgrades.

What is abundantly clear during the initial assessment is that the essence of the Standard is to state with great clarity who manages, performs, verifies, and validates the processes and subprocesses for documentation, implementation, and demonstration of effectiveness.

Endnotes

[1] Many companies prefer to use other terms such as *champion* rather than *steward*. Terms such as *process champion* and *subprocess champion* are used. The thought is equivalent—we must have clearly established and committed leadership.

[2] The assignment of specific elements to the ISO management representative during QMS creation does not relieve the representative of the overall responsibility to coordinate the entire QMS creative process. It is meant to level out the writing and editing load.

[3] See, for example: Defeo, Joseph A., "The Tip of the Iceberg," *Quality Progress*, ASQ, May 2001, p. 29.

[4] See, for example, Neuscheler-Fritsch, Debbie, and Robert Norris, "Capturing Financial Benefits from Six Sigma," *Quality Progress*, ASQ, May 2001, p. 39.

[5] Based on the work of Dr. Anthony F. Costonis, president and founder of Corporate Development Services, Inc., of Lynnfield, MA, at http://www.corpdevelopment.com.

[6] "ISO 9000 Consultants Guide," *Quality Digest*, May 2001, p. 69, at http://www.qualitydigest.com.

[7] See, for example: Hiebler, Robert D., Thomas B. Kelly, and Charles Ketteman, *Best Practices: Building Your Business with Customer-Focused Solutions*, New York: Simon & Schuster, 1998, and Camp, Robert C., *Business Process Benchmarking: Finding and Implementing Best Practices*, Milwaukee, WI: ASQ Quality Press, 1995.

[8] See, for example: Hendricks, Kevin B., and Vinod R. Singhai, "Don't Count TQM Out," *Quality Progress*, April 1999, p. 35, and Tai, Lawrence S., and Zbigniew H. Przasnyski, "Baldrige Award Winners Beat the S&P 500," *Quality Progress*, April 1999, p. 45.

[9] The selection of third-party assessors is integral to the selection of a registrar. For a complete exposition on this topic, refer to Weightman, R.T., "How to Select a Registrar," *Quality Systems Update*, August 1996. Mr. Weightman is the president of Qualified Specialists, Inc., Houston, TX. Also see, Russell, J.P., *The Quality Audit Handbook, Second Edition*, Milwaukee, WI: ASQ Quality Press, 2000.

[10] See, for example: Russell, J.P., "Auditing ISO 9001:2000," *Quality Progress*, July 2001, p. 147, at http://www.asq.org.

[11] The shoulds of the QS-9000 quality system requirements are to be treated the same as the SHALLS of ISO 9001. Should, in this case, indicates a preferred approach. It is not to be confused with the notes of ISO 9001 that are not mandatory, but are used as an interpretive aid.

[12] Schlickman, Jay J., *ISO 9000 Quality Management System Design: Optimal Design Rules for Documentation, Implementation, and System Effectiveness*, Milwaukee, WI: ASQ Quality Press, 1998.

[13] For an extremely lucid discussion of the ISO 9001:2000 and ISO 13485 issue, see Kimmelman, Edward R., "Is ISO Obsolete?" *Medical Device and Diagnostic Industry*, October 2001, p. 76. Mr. Kimmelman is currently the convener of the ISO/TC210, Working Group 1, on quality systems for the medical device industry.

[14] Software development standards include the Carnegie Mellon University Software Engineering Institute capability maturity model for software, which

has become a de facto standard for bids to the Department of Defense and NASA, as well as the IEEE/EIA 12207:Software Life Cycle Processes. All standards of this type can be analyzed and integrated into a QMS using this book's design techniques. See also, Rakitin, Steven R., *Software Verification and Validation: A Practitioner's Guide,* Norwood, MA: Artech House, 1997, p. 7.

[15] Although it is possible to have the registrar declare the organization noncertifiable, I know of no such case in the hundreds of certifications with which I am familiar. The only situation under which this might occur, to my knowledge, is if the facility has obvious safety and/or hazardous waste nonconformances so that the assessors cannot perform their audit in a safe manner.

PART

IV

QMS Effectiveness

But is was the opposite: my father had taught me. Looking at the bird he says, "Do you know what that bird is? It's a brown throated thrush; but in Portuguese it's a ..., in Italian a ...," he says, "in Chinese it's a ..., in Japanese a ...," et cetera. "Now," he says, "you know in all the languages you want to know what the name of that bird is and when you've finished with all that," he says, "you'll know absolutely nothing whatever about the bird. You only know about humans in different places and what they call the bird. Now," he says, "let's look at the bird."

—Richard P. Feynman, The Pleasure of Finding Things Out, Cambridge, MA: Helix Books, Perseus Publishing, 1999, p. 3.

CHAPTER

14

The Biggest Change in ISO 9001:2000 from ISO 9001:1994

To learn something about the Standard, it is necessary to examine the Standard in detail. For example, the Standard mandates that the organization continually improve the QMS effectiveness. Activities that illustrate continual improvement can be obtained by means of such requirements as the quality policy, quality objectives, audit results, analysis of data, corrective and preventive action, and management review (Par. 8.5.1: Continual Improvement of the Standard).

This highly prescriptive language is not only remarkable for a generic Standard but it is backed up by even more prescriptive language in Par. 8.4: Analysis of Data. This mandates that we are to use quantitative methods in the analysis of data, which demonstrates not only how suitable and effective our QMS is but where its effectiveness can be continually improved.

This is powerful stuff! You are required to continually improve the effectiveness of your QMS, and you are required to do it by quantitative means!

Obviously, this is an idealist view of how enterprises actually work. Given the ebb and flow of real-world pressures, an enterprise could not be expected to always show QMS effectiveness. There will often be times when just survival is the primary aim of the enterprise, let alone demonstrating that product return rates have declined. In fact, you may be in the midst of a total recall of your last year's shipments. That anomalous spike in your product return rates would certainly indicate ineffectiveness of your QMS but would not be a predictor of your long-term health.

The Standard should probably have added the caveat, "...and quantitatively indicate how ineffectiveness is being corrected." That is really what is important. If you are doing well, how do you intend to do better? If you are in trouble, how do you intend to dig your way out? In both cases, one should ask how quantitative analysis is used as an improvement tool? This is the answer that I attempt to seek out as a third-party assessor, and I believe this is the intent of the Standard.

The issue of quantitative analysis, by means of effective quality objectives, is discussed in the next chapter.

CHAPTER

15

Quality Objectives

Contents

One of the most powerful methods used to measure QMS effectiveness is to carefully track the organization's progress toward the achievement of its quality objectives. The quality objectives must be written in language that is meaningful to the users and defined by a metric that is uniquely measurable. This does not imply that the metric is easy to measure. For example, if first pass yield is the measurement based on a metric determined by the number of electronic cards that pass final test divided by the total number of cards measured, it is important to make sure that cards that fail the first time, and are subsequently repaired and passed through again, are not included in the first pass data. The application of the appropriate statistics is not a trivial exercise. Use great care in its choice and don't be afraid to change as often as necessary until you have found a truly meaningful metric.

15.1 Quality Objectives Issue

Our experience with 110 ISO 9000–certified companies indicates that the development of quality objectives is one of the most difficult areas of ISO 9000 responsiveness. We see early signs of this issue in our recent ISO 9001:2000 certification activities, where the need for quantitative quality objective expression is mandatory [1].

The primary reason for the observed difficulty appears to be confusion over what constitutes a quality objective and how it should be stated. This observation should not be a surprise because the scope of quality objectives varies widely between small, mid-sized, and large sites. For example, in a small site (less than 50 employees), top management is involved in the day-to-day operations and the president and vice presidents constantly track and analyze performance. By contrast, in a large site (greater than 500 employees) top management requires a number of weekly, monthly, quarterly, and annual reports by managers as a way to track and analyze performance.

In addition, Par. 3.2.5 of ISO 9000:2000 offers a definition of a quality objective in the sense that it is something related to quality that one seeks. Several notes indicate that the quality policy provides the framework for the quality objectives that are intended to flow down through the organization. The definition is so qualitative that it offers a modicum of guidance when we attempt to apply the concept to QMS quality objective design.

It is our purpose, therefore, to present a systematic approach to the design of quality objectives that is flexible enough for use by certified sites of any size. The discussion includes examples of this process related to a small, a mid-sized, and a large certified site. A detailed flowed-down set of quality objectives is presented for a mid-sized site. The use of a Deming cycle is also proposed to ensure the effective implementation of the established quality objectives.

Our approach to quality objectives is global in that we require business objectives and quality objectives to be transparent.

15.2 The Components of a Quality Objective

We begin the discussion with our definition of the five major components of a quality objective:

1. General statement;
2. Metric;
3. Target (goal);
4. Presentation by champion;
5. Flow downs.

The definitions are illustrated by example in Table 15.1. In the table, we indicate how three different-sized sites might respond to three different quality objective statements. The key difference for site size is exemplified in the form of presentation. As the sites increase in size, it is necessary to increase both the number of categories and the frequency of review periods to

Table 15.1
Examples of Quality Objective Components by Site Size

Site Size	Statement	Metric	Target	Presentation by Champion	Flow Downs
Small (<50)	Minimize company's late deliveries	Late shipments divided by all shipments	<1% Late shipments	Production manager graphs percentage by month over all shipments	Impacts purchasing
Midsize (>50 but <500)	Maximize corporate production throughput	First pass yields for each product line	>85% First pass yields	Operations manager graphs percentage by month by product lines	Impacts production assembly and test and purchasing
Large (>500)	Maximize divisional proposal wins	Bids won divided by total bids (by division)	>40% Winning proposals.	Vice president of sales graphs percentage by week by division	Impacts divisional marketing and sales

effectively track the larger set of data. For example, for a small site, all the shipments are counted, independent of product type, because the number of customers is limited—it could be just one—whereas for the mid-sized site, the graphs are plotted for each product line.

For each quality objective statement, there is a clearly defined metric, target, form of presentation, and a statement with regard to flow down. For each impact, a subsidiary quality objective statement is required. Table 15.2 illustrates typical flow-down impacts.

In Table 15.2, we indicate how each primary objective that has been established by top management is assigned a metric, target, champion, method of presentation (on an intranet), and what the flow-down objectives could look like [2]. The exact number of flow-down levels is highly dependent on the site size. The flow-down objectives fulfill the Standard's requirements to do the following:

- Meet requirements for product;
- Be established at relevant functions and levels within the organization;
- Be measurable.

The table considers manufacturing's primary quality objective: "Ship product as specified by the customer-agreed-to shipping date." Then, the impact of other departments that are essential to the successful achievement of this objective are considered, and objectives are established for those departments in a way that supplements manufacturing's efforts. The flow down carries

Table 15.2
An Example of Flow-Down Quality Metrics

Manufacturing Primary and Supporting Quality Objective(s)			
Primary objective: ship product as specified by the customer-agreed-to shipping date			
Metric	Target	Champion	Intranet location
Percentage of shipments that meet ship date	>95%	Vice president of manufacturing	Shipping.xls
First support objective: reduce NCMRs in assembly			
Metric	Target	Champion	Intranet location
Number of NCMRs per product line	Zero	Assembly supervisors	NCMRs.xls
Second Support objective: optimize first pass yields			
Metric	Target	Champion	Intranet location
First pass yields per product line	80%	Vice president of manufacturing	Product Yields.xls
Third support objective: optimize vendor/subcontractor evaluation on-time deliveries			
Metric	Target	Champion	Intranet location
Vendor percentage on-time deliveries	>98%	Purchasing supervisor	VendorsOT.xls
Fourth support objective: optimize response to nonconformities			
Metric	Target	Champion	Intranet location
Response time to resolve nonconformities	Minimize	Vice president of quality assurance	NCRTime.xls

through four stages and involves the participation of assembly supervisors, the purchasing supervisor, and the vice president of quality assurance in support of the vice president of manufacturing.

The quality objectives must also be consistent with the quality policy. If we assume that the quality policy statement is, for example, "Quality within the Excellent Corporation means never being satisfied with anything less than a delighted customer," we easily meet this consideration. You will need to make sure that your quality objectives are in harmony with your quality policy statement.

It is important to keep in mind that to create an effective QMS, the quality objectives must be written in language that is meaningful to the users and defined by a metric that is uniquely measurable. This does not imply that the metric is easy to measure.

For example, if the on-time-delivery metric is determined by the delivery date to a customer (e.g., less than 30 days from receipt of order), you are subject to traffic and docking issues that are beyond your control. By contrast, if the shipping date determines the delivery time, you have total control of the measurement. In addition, for this metric, it is necessary to have at least two distinct categories: (1) shipping dates for product that the customer has not revised the schedule, either verbally or by written notice; and (2) a category in that numerous revisions have been allowed. Otherwise, the measurement will be meaningless.

The application of the appropriate statistics requires a careful evaluation of what makes sense for the site. Do not be afraid to change the measurement method as often as necessary until you have found a truly meaningful metric.

15.3 The Framework for Quality Objectives

Quality objectives are an integral part of the QMS design framework. The position of the quality objectives in the total QMS framework is illustrated in Table 15.3. As indicated, each component of the QMS imperatives flows down to the next level in a continuous movement. The order of presentation begins

Table 15.3
Location of Quality Objectives Within the ISO 9001:2000 QMS Site Imperatives

Site Imperatives	Typical ISO 9001:2000 Quality Manual Paragraph Content
Business and certification scope (requires registrar's review and acceptance)	What are the services and products that are under the certification? Where is the organization located? Is it multidivisional with multiple sites?
Vision statement	What is the long-term objective of the organization? What dominance does the organization wish to have long term?
Mission statement can be corporate level or divisional/department level	What are the key objectives the organization needs to achieve midterm to achieve such dominance? Which methods are appropriate to this purpose?
Quality policy statement	What is the quality policy statement that is easy for every employee to remember and that clearly defines how we are to treat our customers?
Quality objectives/metrics	Define the quality objectives that lead to continual improvement, which is measured, for example, through the trend analysis of (a) customer satisfaction and dissatisfaction (customer returns and complaints, reorders, overall market share); (b) internal improvement metrics (yields, scrap); (c) corrective and preventive actions; and (d) return on net worth.
Process-based QMS	Define the organization's core competencies and clearly define the methods used to provide adequate resources to allow fulfillment of the quality objectives.
Customer needs and expectations	What are the ways and methods used to clearly define customer needs and expectations? How do we determine customer perception of our services?
Propagation of the quality policy	What are the methods used to make sure that all employees, suppliers, and customers understand our quality policy, quality objectives, and status of our progress to achieve those objectives?

with a strategic statement of the overall site's scope of certification and industrial position. From this framework, the vision, mission, quality policy statement, and then quality objectives are established. For completeness, the table includes typical ISO 9001:2000 quality policy statements that could be included as part of the quality manual text. The site's complete strategic framework also includes its process-based QMS, the manner in which customer needs and expectations are fulfilled, and the manner in which the quality policy is propagated throughout the site [3].

15.4 Universal Quality Objectives Process

From this model of imperatives and the examples presented, we can derive a universal approach to quality objective design. This universal process is illustrated in Figure 15.1. In this concept, we form cross-functional action teams to expedite the creation, measurement, and analysis of quality objectives. For our example, the cross-functional teams are defined as objectives action teams (OATs). They, of course, could be called whatever is commonly used within an organization. The OATs follow the classic Deming cycle illustrated in Figure 15.2. Based on the defined components of a quality objective, the teams determine and ensure that quality-objective implementation is performed effectively. This type of approach to quality objectives design has been used successfully on both ISO 9000 and TQM implemented programs [4].

We have found that the creation of effective quality objectives has been an issue since the inception of the ISO 9000 schema in 1987. We have proposed a clearly defined quality objective process with examples for ISO 9001:2000 certification teams to use, which we believe can greatly enhance the value of their quality-objectives process. The ideas presented have been used successfully in the creation, implementation, and analysis of quality objectives applied to both ISO 9000 and TQM programs.

The exact manner in which the data is analyzed, plotted, and presented to top management is a matter of what makes sense for your organization. What is generally overlooked is that statistical techniques includes nonstatistical techniques (i.e., the manner in which data is represented in terms of Pareto charts, run charts, histograms, and other common forms of analytical display is as much a part of the statistical concept as SPC), which may or may not be useful.

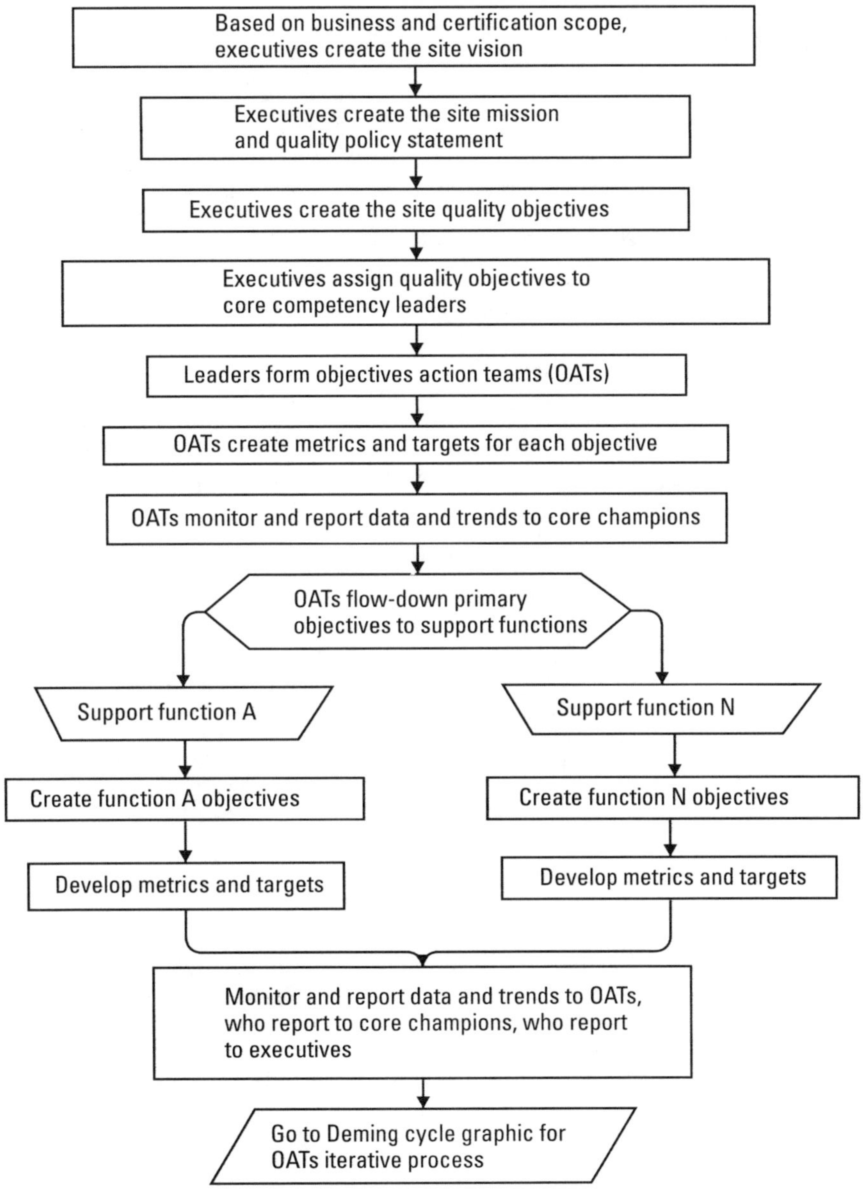

Figure 15.1 Universal quality objectives design process.

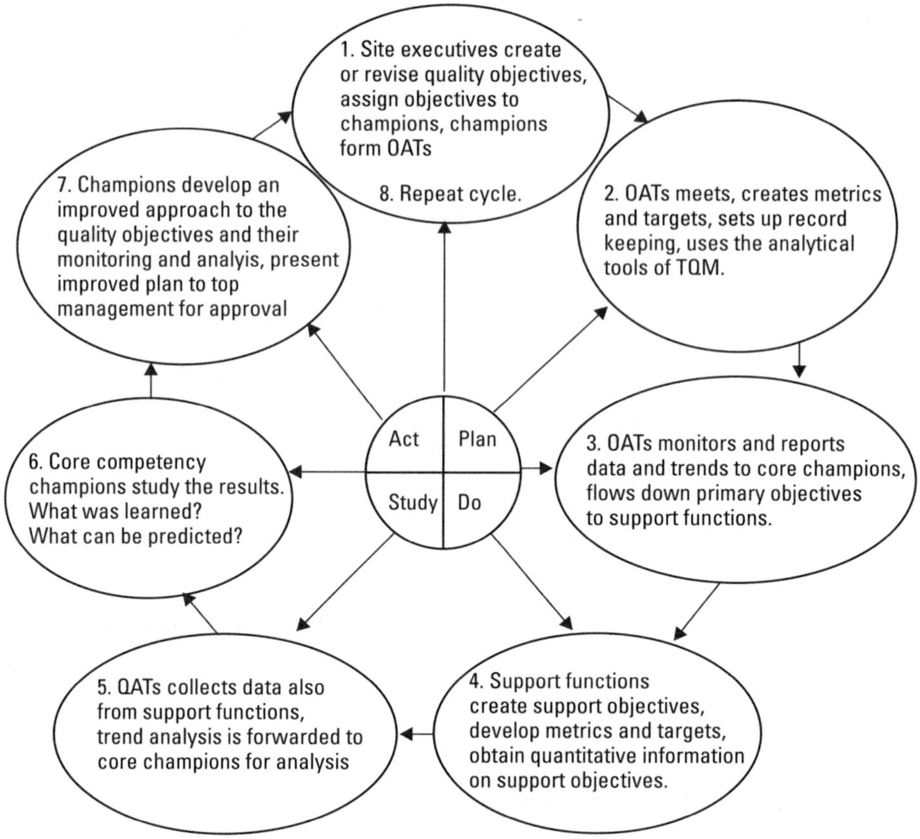

Figure 15.2 Application of the PDSA Deming cycle to ISO 9001:2000 quality objective design.

Endnotes

[1] The general issues inherent in setting up a QMS compliant with the Standard have been addressed in such articles as Ketola, Jeanne, and Kathy Roberts, "Transition Planning for ISO 9001:2000," *Quality Digest,* March 2001, p. 24, and McLymount, Rosalind, and Amy Zuckerman, "Slipping into ISO 9000:2000," *Quality Digest,* August 2001, p. 30, at http://www.qualitydigest .com.

[2] Online quality management systems have become common, especially for tier I and tier II documentation. See, for example, Accardi, Valeri, "ISO 9000 Without a Paper Manual," *Quality Progress,* November 2001, p. 86, at http:// www.asq.org.

[3] See, for example: Cianfrani, Charles A., Joseph J. Tsiakals, and John E. (Jack) West, *The ASQ ISO 9000:2000 Handbook,* ASQ Quality Press, 2002, at http:// qualitypress.asq.org.

[4] TQM methods are based on the work of Dr. Anthony F. Costonis, president and founder of Corporate Development Services, Inc., of Lynnfield, MA, at http://www.Corpdevelopment.com.

PART

V

QMS Styles

In physics, the interpretation of experiments are models or theories, and the realization that all models and theories are approximate is basic to modern scientific research. Thus the aphorism of Einstein, "As far as the laws of mathematics refer to reality, they are not certain; and as far as they are certain, they do not refer to reality." Physicists know that their methods of analysis and logical reasoning can never explain the whole realm of natural phenomenon at once, and so they single out a certain group of phenomena and try to build a model to describe this group. In doing so, they neglect other phenomena and the model will therefore not give a complete description of the real situation.

—Fritjof Capra, The Tao of Physics, New York: Bantam Books, 1984, p. 27.

CHAPTER

16

Contents

Readership and Form

16.1 Which Comes First? The Manual, the Processes, or the Procedures?

We want to make sure that our QMS creation model is complete. For example, we have seen that the creation of the ISO 9000 documentation system is an iterative process whereby each document tends to support other documents. As a result, the question arises as to which document to create first.

There are several ways to approach this question, and all three approaches will have some percentage of the others in practice:

- *Top-down method*—Begin with the manual's quality policy statements and then create the lower tier documents.
- *Bottom-up method*—Work from the set of lower tier documents to create the quality policy statements in the manual.
- *Process-flow method*—Start with a flow analysis of the organization's processes and create the manual and lower tier documents concurrently. The methods used to create the flow analysis include flow charts and a tabular flow of activities. This is an iterative process. However, the flow analysis ensures consistency between the various documentation levels and minimizes redundancy, especially between the policy statements and the procedures.

Alert: If flow charts are chosen for the flow analysis, be sure that there is compatibility among the computer systems and that enough expertise is available to program the flow charts and modify them.

Table 16.1 indicates which method might be the most effective for a given size organization with a given degree of documentation maturity.

We have observed all of these methods in practice to some degree and believe that the most effective approach overall is to begin with an analysis of the organization's processes first (i.e., the process-flow method) [1].

16.2 Par. 4.2.1 of the Standard

The Standard does imply a form of style. This is found in Par. 4.2.1: General Documentation Requirements, Note 2, which advises us that QMS documentation can vary considerably between organizations due to differences in size, complexity, and personnel, among others. This is a very powerful note because it implies that the Standard allows the organization to custom fit the documentation to the organization's sophistication and complexity. Certainly, a design engineer requires more guideline than procedure, whereas an assembler requires more procedure than guideline.

Although it seems intuitively obvious, it's worth stating the following:

- Design engineers require primarily guidelines.

Table 16.1
Comparison of Documentation System Methods

Method	Suggested Applications	Comments
Top-down Manual ↓ Lower-tier documents	Small organizations with minimal documentation, such as Metrology Laboratory or Component Repair Co.	Tendency to stress policy over procedure—but this does not have a great impact because on-the-job-training is prevalent
Bottom-up Manual ↑ Lower-tier documents	Large organizations with readily available mature documentation systems that were originally based on Mil-Q-9858A or standards	Little opportunity to streamline the existing system Redundancy and obsolescence needs to be looked at in detail
Process-flow Manual→→↓← ↑ ↑ ↓ ↑ Process ↓ ↑ ↓ ↓ ↑ Lower-tier documents	Appropriate for all types of organizations Generally by means of a flow chart or flow table Found to be the most effective method for explicit organizational knowledge	Tends to minimize total document quantity Exposes weaknesses in operational activities Readily catches the most frequently found gap—the handshake from department to department Most expeditious for ISO 9001:2000 manual creation

- Experienced machinists need a minimum of procedures.
- Test operators need more detailed process sheets to properly perform their functions.

It is common, for example, to find engineering documents at a level of detail that is literally overwhelming. This issue can be effectively resolved by use of a controlled one-page flow diagram that references the key design documents. This becomes a handy wall reference chart for the project engineer. It's also common to find tier II and related tier III documents that contain nearly 50% of the same text. However, on the first final edit, it is better to be cautious so that the baby is not thrown out with the bath water. We maintain

A little long,
May be too strong,
But it ain't wrong.
Terse is worst!

16.2.1 Linear Estimate

At the risk of oversimplification, we offer a rule with regard to necessary detail (see Figure 16.1). The rule assumes a linear function that is certainly not what you normally find, but it should clarify the point.

Notice that more training is usually required as the subtlety of the documentation context expands (e.g., the ability to understand work instructions versus policy statements). The more expert the reader, the more guidelines make sense. The less expert the reader, the more highly detailed work instructions make sense.

There is a flip side to this concept. Work instructions are always required, no matter what the reader's competence level, when it comes to detailed test or laboratory instructions.

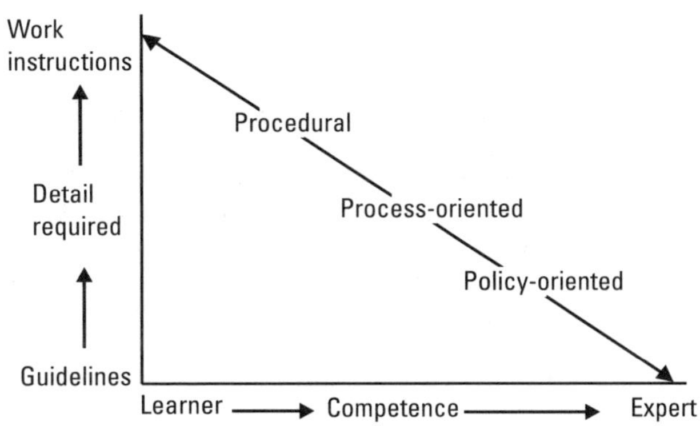

Figure 16.1 Documentation rule to "consider the reader."

16.2.2 Conclusion

We do not imply that everyone needs to be a great writer—it is unclear how to define such a concept. But it is within each author's ability to stress clarity and to simplify concepts. If your text looks long and cluttered, it is. If your key ideas do not show up against the background of words, they are lost. Clear exposition is based on some basic rules, but it is also based on intuition and common sense.

A way to produce an effective document is as follows:

1. Start with one rough draft.
2. Undergo two subsequent editorial reviews with the document users before document control is initiated.
3. Going to document control too early is an enormous waste of time and energy. Later is better.
4. Some companies do not place their documents under control until just prior to the readiness (preassessment) audit.

For most companies, there will exist a strong tier III and tier IV set of documents already available, especially if they have a basic quality system in place based on a TQM or Food and Drug Administration–driven set of requirements. The biggest issue is to integrate the tier I and tier II documents into this pool of readily available documents and then fine tune the entire set to make them consistent and effective.

Our observations in well over 100 organizations have indicated that when the manual is written for the customer, and especially for the new customer—in a clear, concise manner, filled with specific information for decision makers—it is an effective document for all other readers.

Endnote

[1] The reader may feel that there is a conflict between the use of process analysis as the first element of system design and the belief that the first major gate in system development is the manual. We maintain that when we begin with process and then use that information as a database to define our quality policy statements, the policies end up with greater credulity and usefulness. The process documents act as the baseline research tools and the quality policy manual acts as the system controller. There appears to be some agreement in this area (i.e., McLymont, Rosalind, and Amy Zukerman, "Slipping into ISO 9000:2000," *Quality Digest*, August 2001, p. 30, at http://www.qualitydigest.com.

CHAPTER

17

Contents

The Adverse Effects of Paraphrasing

Let us just say that there are two sorts of poetical minds—one kind apt at inventing fables, and the other disposed to believe them.
—Galileo Galilei, Dialogue Concerning the Two Chief World Systems, Translated by Stillman Drake, New York: The Modern Library Science Series, 2001, p. 488.

17.1 The Two Classes of Paraphrasing

17.1.1 The Issue

We have set aside a full chapter to deal with paraphrasing because in our experience paraphrasing trivializes the Standard—although those who tend to paraphrase believe it to be a form of simplification (i.e., minimalism). We intend to show that it is much more a form of nihilism (i.e., repudiation). For example, it is interesting to note that in those organizations where the manual has been highly paraphrased, the tendency is not to audit the manual during internal quality audits. Auditors know when something isn't worth their time and effort. In those cases, of course, the manual falls behind in its currency and relationship to enterprise reality (e.g., out-of-date organizational charts, processes, and quality objectives).

17.1.2 Classes

There are typically two classes of paraphrasing that we define as follows:

- Class I—a direct restatement of the Standard with minor modifications;
- Class II—a table of contents list of where to find information in the lower level documents based on a direct restatement of the Standard.

17.2 Paraphrased Class I Characteristics

We begin with a discussion of class I paraphrasing. The best way to define this issue is to give an example of a paraphrased quality policy statement. We have chosen the Standard's requirement, 8.2.2: Internal Audit.

The following is a typical direct paraphrasing of this requirement (essentially word by word of the Standard's text):

> Excellent's managers who are responsible for the area being audited shall ensure that actions are taken without due delay to eliminate detected nonconformities and their causes. Follow-up activities shall include the verification of the actions taken and the reporting of verification results.

17.2.1 ISO 10013:1995

By contrast, the ISO 9000 Guideline on Quality Manuals, ISO 10013:1995, gives an example of how to respond to a descriptive quality policy statement with prescriptive statements with regard to internal audits. The 1994 requirements were very similar to the 2000 requirements. The difference in language and intent from paraphrasing is obvious. Each SHALL is broken out into prescriptive statements that discuss such items as function management, authority and responsibility for audit reports, and management review of the audit results. Because of the clarity and specificity of the ISO 10013 quality policy statements, people have argued with me that what was presented was actually a procedure—even though the guideline was absolutely clear that it was an example of a quality manual section. The confusion in this area is mind boggling.

17.2.2 Discussion of the Direct Method of Paraphrasing —Class I

In the direct method of paraphrasing, all or nearly all of the Standard's text is used as a quality policy statement. As a result, a manual written in this fashion

- Looks, reads, and feels like the Standard itself;
- Has little to differentiate the text from that of a competitor;
- Fails to define the prescriptive rules of the house.

For example, a purchasing or quality-assurance manager who receives such a manual during a make-buy decision would have little information to go on. Any employee who read the manual would be hard put to understand the dynamics of the organization and its commitment to ISO 9001:2000.

Also, if we compare the direct method of paraphrasing with the ISO 10013 Guideline (i.e., Table 17.1), we see that the contrast is significant in terms of information transfer and clarity. The response in ISO 10013 offers a look into the actual operation of the company while the paraphrased text could be written about any number of competitive organizations. The competitive advantage is negated. Moreover, the paraphrased text maintains the future tense, so it is not clear if this is what is going on now or later. In addition, there is nebulous information in the paraphrased text so any decision maker would find it difficult to decide on the depth of quality in that company.

17.3 Paraphrased Class II Characteristics

Another technique commonly used to paraphrase is to put the language into a table of contents format. A typical pattern for 7.6: Control of Monitoring and Measuring Devices is as follows:

Table 17.1
Attribute Comparison of ISO 10013 Versus Direct Paraphrasing

Attribute	ISO 10013	Directly Paraphrased
SHALL response	Each SHALL responded to with a quality policy statement	Each SHALL restated without a prescriptive response
Clarity/tense	Simple declarative sentences in the present tense	Restatement in the future tense—questions arise as to whether or not the action has happened yet
Detail	Sufficient for decision makers to make judgments about the efficacy of the QMS and to prepare an audit plan	Nebulous information—restates the Standard
Responsibility	Clearly stated	Does imply some responsibility
Market differentiation	Contains the personality and pulse of the organization	Looks and sounds like everybody else
Reference to procedures	Directly stated as Document QA 123-4	In general, also directly stated
Configuration	Directly sequenced to Standard's elements	Directly sequenced to Standard's elements

Paraphrased example

The Excellent Corporation's control of monitoring and measuring devices procedure demonstrates that we do the following (refer to Doc. # MET-2-005):

- Determine the measurements to be made and accuracy required;
- Identify all monitoring and measuring devices that can affect product quality;
- Define the process employed for the calibration of monitoring and measuring devices;
- Identify monitoring and measuring devices with suitable indicators or approved ID records;
- Maintain calibration records;
- Assess and document the validity of previous inspection and test results when monitoring and measuring devices are found to be out of calibration;
- Etc, etc, etc.

17.3.1 Discussion of the TOC Approach to Paraphrasing—Class II

The table of contents approach at first glance almost sounds and looks like a prescriptive set of quality policy statements. If we rewrite a few of the bullets, it will become easier to see the difference when responsive statements are actually used.

Recommended Quality Policy Statement Response

- The identification, calibration, and adjustment of all equipment at the Excellent Corporation are the responsibility of operations engineering. Calibration plans are managed via logs that are maintained to indicate calibration cycles and frequency.
- Calibration labels are used on all test, measurement, and inspection equipment to alert operators that calibration is adequate or due. If calibration is overdue, operators are to immediately alert operations engineering and suspend using the equipment until calibration is completed.
- All equipment is sent out for independent calibration to companies selected on their capability with regard to using nationally known standards. Operations engineering maintains logs of all of these transactions.

A Paradox database file, CALIBRAT.DB, is maintained listing calibration status for all equipment on a calibration cycle.

- Etc, etc, etc.

17.3.2 Comment

Obviously, the feel is very different in the nonparaphrased response, and the information transmitted is far more useful to any reader of this manual. The rules of the house are clearly stated with respect to monitoring and measuring devices.

A quality-assurance manager who was interested in gauging the technical depth and metrological competency of the Excellent Corporation would be favorably impressed. In actual practice, they are deeply impressed.

It is important to note that this response requires that the section be written by those who are experts in the monitoring and measuring area. An experienced third-party assessor can tell within minutes the manual's authorship and whether it is technically sound. For example, a manual that has been written by the quality-assurance manager has one voice. The sections dealing with quality-assurance issues are interesting and informative. The rest of the manual can sound hollow and more like a summary of the Standard.

17.4 Conclusions

As demonstrated, we believe that the use of paraphrasing (restatement of the Standard) is a matter of one's decision to obscure rather than to clarify. The paraphrased approach does the following:

- Trivializes the Standard;
- Minimizes the opportunity to question the organization's processes and thereby improve them.

Such trivialization is anathema to the purpose of ISO 9000—that is to support continuous/continual improvement. In fact, in the case where one individual writes the manual and also paraphrase's the Standard, the loss of clarity is magnified even further. Yet, it is our experience that such events frequently occur.

Therefore, although we have found paraphrasing often used, we feel that paraphrasing is inherently an ineffectual and inappropriate method of communication. Unfortunately, paraphrasing is widely encouraged by consultants and tolerated by registrars.

We maintain the following about paraphrasing:

- In its restatement of already known phrases of the Standard, it offers the reader minimal organizational information and obscures the uniqueness of the quality management system—it is out of phase with the concept of information technology flow.
- It fails to clearly instruct the executive staff of the importance of continuous improvement and the manner in which the staff is to achieve this goal.
- It is clearly not in the spirit of ISO 10013:1995 (p. 11), which demonstrates that enough detail is necessary to paint an operational picture of the organization's response to a given requirement.
- It does not permit the reader to grasp the organization's technical and manufacturing personality.
- It negates the marketing and sales potential of the manual—it makes every organization sound like every other organization, and we are befuddled as to why any organization would want to be seen as undifferentiated from its competitors.
- It is a form of intellectual dishonesty because it trivializes the intent of the Standard, which is to provide a clearly stated and definitive top-down, executive view of the organization.
- It has been found to be a tool used primarily by a single author instead of a group of technical experts, thereby diluting the technical integrity of the manual.
- It makes the manual so useless in the eyes of internal auditors that they don't even bother to audit it.
- It makes the manual so useless in the eyes of the customer that they cannot use it as a reliable tool in purchasing decision making.

Third-Party Impact We demonstrated several approaches to paraphrasing and indicated the difficulties inherent in such an approach from a general viewpoint. We have also observed that usually in the case of an accredited third-party assessment:

- Direct paraphrasing is not acceptable because it cannot be clearly audited. It forces the auditor to make assumptions regarding the policies of the organization. They are not permitted to do so.

- The table of content technique is also not acceptable for the same reason, as the assessor is still forced to assume what the policies (house rules) are.
- Paraphrasing forces a compromise position that dilutes the audit value to the client—the assessor is forced to search the lower tier documents to find the quality policy statements. Result: less bang for your audit buck!

Alternatively, however, there are consultants, registrars, and assessors who do not agree with this conclusion (some vigorously), and if the reader feels the same, you should be able to find suitable partners for your certification effort.

Obviously, we consider paraphrasing unacceptable in any manual. The formation of informative quality policy statements will force you to better understand your QMS and will orient your intellectual efforts in the direction of continual, enterprise improvement.

CHAPTER

18

Contents

Publication Media

18.1 Selection of a Publication Media (Hard-Copy Versus Electronic)

The use of a myriad selection of software applications such as enterprise resource planning, customer relationship management, and supply chain management software, which utilize the Internet and intranet environments, are commonly seen in both small and large enterprises. Web-centric systems provide powerful integration techniques to mold engineering, supply-chain management, manufacturing, and planning into networks known as collaborative product commerce systems [1].

In addition to the overall global gains in efficiency, speed, and productivity in sales and marketing, engineering, and manufacturing, the Internet and intranet is now used to solve what had previously been sticky issues with regard to documentation distribution and control of satellite sales offices and customer service field operations.

Of course, with software comes reliability and interoperability issues that are far beyond the scope of this book. The software division of the ASQ, for example, publishes extremely perceptive articles on this subject [2].

18.1.1 Media Types

As a result, it is vital to select the publication media type (i.e., hard copy, electronic [online], or a mixture) that best suits the organization's capabilities. The documents must be controlled in some fashion. The following are some examples of control:

- Stamped;
- Controlled versus uncontrolled check marks;
- Different colored icons or strips or pages;
- Page count noted;
- Provisions made for the issuance of uncontrolled manuals (e.g., to staff, employees, customers/clients, and subcontractors);
- Provisions made for the issuance of hard copy in online systems for training, revision control, and the use of hard-copy documents for specified limits of time by production personnel.

18.1.2 What Should Be the Exact Form of the Documentation System?

The answer of course will depend upon the specific needs of the organization. However, we can examine a generalized scenario and then look at a few specific cases to give the reader an idea of how this type of analysis is carried out. Once a particular alternative is chosen, it will be necessary to immediately begin to train all employees on that protocol. There is no easy answer to this question; it is a matter of experimentation to determine which system best fits our objectives.

A good rule is to figure that it will take twice as long to train people on online documents than on hard copy because of searching difficulties in systems with multiple directories. Very few employees will be comfortable when they are asked to find a document that they do not normally use in an online system. This is why it is a good design rule to set up the system with hub documents that start with an easy to find entrance icon, which sends the employee to the main ISO 9001:2000 directory and from there to a set of hub documents.

18.1.3 Control Issue

We begin with a simplified decision matrix portraying control versus documentation (see Table 18.1). We examine the possibilities for either central or local area manager (LAM) document control versus the type of document (i.e., either the manual or lower tier documents). Although there is another protocol—the concept of a local area user (LAU), whereby the user becomes responsible for the use of the correct document revision level—this approach is difficult to manage and is not covered further in our discussion, other than to say that each user must be effectively trained in document control protocols.

Table 18.1
Decision Matrix for Documentation Systems

Control	Type of Documents	
	Quality manual	*Lower tiers*
Centrally controlled	Online	Online
	Hard-copy	Hard-copy
	Hybrid	Hybrid
LAM controlled	Nonapplicable	Online
		Hard-copy

The chosen set of protocols includes online, hard copy, or a mixed (hybrid) system. A hard-copy system has another degree of complexity in that the documents can be distributed as either do the following:

- A set in binders;
- Individual documents.

18.1.4 An Example of How to Choose What Is Best for You

To limit the number of choices, we assume that your organization is characterized by the following:

- *Little or no central control.* A central document control center does not exist, nor does it readily fit into the economic viability of the organization. Top management would agree to the formation of such an organization, but under duress.
- *Local area managers.* These might be department heads who will agree to maintain document control procedures that include locally controlled master documentation lists if it is the best way to go.
- *A networked system.* Such a system should be readily available so that tiers I and II can go online immediately. There are not enough terminals for tier III documents to be used in production.

Therefore, of all the many possible configurations, it appears that we only have two practical choices:

1. Set up tier I and II to be centrally controlled and have tiers III and IV controlled by LAMs. Pro: This way is the easiest for finding high-level

documents and the network is readily available. Con: This way requires a dedicated central document control manager and is more expensive.

2. Set up tier I central and have tiers II, III, and IV controlled by LAMs. Pro: This way requires minimal central document control management. Con: This way is difficult for finding high-level documents, but the LAMs are willing to work with this documentation control structure.

In an actual case, alternative I was chosen. The difficulty in finding high-level documents turned out to be the decisive factor. The central document control function was shared by several employees and consumed a modest amount of time once the system was in maintenance.

18.2 Generic Numbering System

In many cases, the documentation system is a mixture of online and hard copy. This raises the issue of just what type of numbering system will work concurrently. One possible approach is to have the online format as shown in Table 18.2.

In this table, we see the following:

- Only eight integers or alphas are needed.
- It permits 99 tier III documents for every tier II document.
- The department and tier II levels do not require a number because they are in computer directories and subdirectories.
- The level number is either 1, 2, 3, or 4 for the various tiers.
- The element of the Standard would run from 004 to 853.
- The document number runs from 01 through 99.
- Revision control runs up to 99.

Table 18.2
Numbering System for Online Format

Dept	Tier II	2	852	12	01
(Directory)	(Subdirectory)	(Level)	(Element of Standard)	(Document)	(Revision level)

When hard copy documents are required, we would simply add the operating departments code name. For example, quality assurance (QA), and the type of document (CAP for corrective action procedure) would form QACAP28521201 = the first revision of the QA tier II corrective action procedure, document number 12.

Alternative to Numbered Online Systems Because of the use of hyperlinks and clearly defined subdirectories, it is common to find online systems that do not use a numbering system at all but simply link to titled documents. Thus, the system would simply link you to the corrective and preventive action process document, which resides in a subdirectory dedicated to quality-assurance documents. The use of numbering systems is generally found with documentation systems that have been in use for many years, particularly in aerospace, military, and medical environments.

Endnotes

[1] Vogel, Steven A., "Satisfying QSR Requirements with Collaborative Production Management Systems," *Medical Device and Diagnostics Industry*, July 2001, p. 90, at http://devicelink.com/mddi.

[2] See, for example, Kress, Michael P., "An Approach to Harmonizing ISO 9001:2000; AS9100:2001; and ISO 9000-3 for Software," *Software Quality*, Fall 2001, p. 10.

CHAPTER

19

Contents

Writing Style

The manual, in order to satisfy such a diverse audience and to meet the needs of its customers, should follow several style guidelines [1].

19.1 Contain Paragraphs and Sentences That Are Variable in Length, but Short

- One idea per paragraph.
- Realize that you will normally write about 40% more than is necessary. It will take about two years after your certification to clean up the documents so that they are effective.
- Stream-of-consciousness writing can be interesting in novels, but it is onerous in technical material.

19.2 Use Simple Declarative Sentences

Examples:

- The planning function at Excellent is the responsibility of production control.
- Production control supplies the materials department with a daily list of raw material requirements by means of the MRP system.

19.3 Avoid Redundancy~~, i.e., repeated material~~

- For example, if the manual contains all of the quality policy statements, there is no need to restate them in lower tier documents—a common form of redundancy.
- Redundancy confuses the reader and forces the assessor to compare redundant sentences. Invariably they will differ and you may receive a nonconformance based on the degree of difference. The classic places for redundancy occur between tier II and tier III documents, and between flow charts and their attendant text, which usually just repeats the flow chart information.

19.4 Stress the Active Voice (Subject, Verb, Object)

- Preferred: The president has designated the director of quality assurance as the ISO management representative,
- Avoid: The ISO management representative has been designated by the president as the ISO management representative.

19.5 Clearly Label Section Content

It is a good idea to outline your work before you start. In that way, your work is initially structured for clarity. The structure will then drive the correct labeling. Of course, the outline is alive and will change on you—be prepared to suffer.

19.6 Build a Useful Table of Contents (TOC)

Be sure to indicate the relationship between the TOC sections and the specific clauses of the Standard at the highest level of the Standard as possible.

For example, if Section 4 of the manual is termed *Quality Management System* and deals with all the SHALLS of the Standard's 4.0 requirements, you need only reference Section 4 to the Standard's 4.0 (see Table 19.1).

However, if the format is such that quality policy is in Section 2 of the manual, it will be necessary to reference that specific paragraph to Par. 5.3 of the Standard (see Table 19.2).

Table 19.1
Example of a Directly Referenced Manual TOC

Manual Section	Section Title	Standard Reference Element Number
1	Scope of the QMS	1.0
2	History of the Enterprise	—
3	Organization Vision and Mission	—
4	Quality Management System	4.0
5	Management Responsibility	5.0
6	Resource Management	6.0
7	Product Realization	7.0
8	Measurement, Analysis, and Improvement	8.0

Table 19.2
Example of an Indirectly Referenced Manual TOC

Manual Section	Section Title	Standard Reference Clause Number
1	Quality Manual	4.2.2
2	Quality Policy	5.3
3	General Requirements	4.1
4	Management Commitment	5.1
...	...	....
18	Corrective Action	8.5.2

19.7 Minimize Organizational Jargon, but Keep the Industry Language

Acronyms such as CEO, COO, and CFO are fairly well recognized internationally. However, short forms like DQA (director of quality assurance), and DCA (document control administrator) that may or may not be familiar within the organization (and I will tell you from experience they seldom are) place a burden on the reader—that in most cases will simply turn them off.

As big a pain in the neck as it is, it is far better to spell out the titles every time than to rely on the reader's memory. It will be appreciated. However, do not throw away the language of the industry. The manual is read by experts—write to their level.

19.8 Write To Be Understood, Not to Impress

Contrary to the opinion of many, stream-of-consciousness technical writing is a quick way to lose your reader. Just think about the last time you tried to read a specification sheet that used 100-word sentences. They certainly are impressive, but they have minimum affective value.

19.9 Clearly Define Terms

It is far better to define the terms in the text as they appear. However, a glossary is an effective back-up approach. It will be used from time to time, believe it or not.

19.10 Effectively Link the Reader to Referenced Documents

Once you have the reader's attention, you want to keep it. A clear reference to the associated lower tier document maintains the interest. There are a number of ways to link, for example:

- *Direct reference:* "The marketing and sales process is described in Doc# 7-2-001-0206, entitled 'Standard Operating Procedure for Marketing and Sales.'"
- *Indirect reference:* "The lower tier process documents for marketing and sales are listed in Appendix A, entitled 'Master List (or Document Tree) of Lower Tier Documents by Manual Section.' The related documents are found under the column (tree) denoted as marketing and sales."
- *Hyperlink:* "Please use the icon entitled 'Design Control Process' for further information."
- *Hybrid systems:* Systems that contain both electronic and hard-copy files are the most common documentation systems in use. It is imperative to clearly define which documents and records are online and which are hard copy. Most importantly, it is vital to clearly define if electronic signatures are in use, how the software is validated, and how the electronic signatures are protected. For example, this is a critical requirement for medical devices.

19.11 Use Bullets or Equivalent Symbols Wherever Possible

Technical information overload occurs much faster than one would think, even among those immersed in the subject [2].

19.12 Avoid Words That End in "ing"

Preferred: Presidential responsibilities include the following:

- Assign the ISO management representative;
- Chair the management review;
- Approve the business plan.

Avoid: The duties and responsibilities of the president, including review, are the following:

- Assigning the ISO management representative;
- Chairing the management review meetings;
- Approving the business plan.

19.13 Use the Spell Checker, and Then Don't Believe It

This is because manger and manager are both okay with the spell checker. So are know and now, he and the, and through and thorough.

19.14 Use Graphics Whenever Possible for Tables, Figures, and Flow Charts

This is especially true for online systems, where only a portion of the total document can usually be seen at a time. It would be nice if all documentation systems were in the form of portrait displays on newspaper-sized screens showing multiple pages—but not today.

19.15 Avoid the Future Tense—Stay with the Present Tense

Avoid the future tense. Either it is happening or it is not. The use of SHALL and will leaves the issue hanging and cannot be readily audited. SHALL is used in the Standard because it is a future requirement of the organization (you). Once the organization has responded to the SHALL, it is now in the present.

As an example, instead of "Every department manager shall hold a monthly quality review session with their staff," we prefer "Each department manager holds monthly quality review sessions with his or her staff."

Of course, if you do want to include a future event, the future tense is appropriate (e.g., "In 2003, the present tracking system will be replaced with an MRP system").

Endnotes

[1] There are dozens of good books on clear writing. I have often been asked why I haven't read them. One very nice little one is Bates, Jefferson D., *Writing with Precision,* Washington, DC: Acropolis Books, 1985. The old reliable one is Crews, Frederick, *The Random House Handbook,* New York: Random House, 1974. The issues of manual style are also addressed in publications such as Russo, C.W. Russ, "12 Rules To Make Your ISO 9000 Documentation Simple and Easy To Use," *Quality Progress,* March 1997, p. 51.

[2] The seminal work of Robert E. Horn of Information Mapping, Inc., Waltham, MA, has demonstrated that the effective communication of technical concepts requires a rate of approximately three to five ideas at a time as a specific block of information.

PART VI

QMS Design Rule Summary

We are at the very beginning of time for the human race. It is not unreasonable that we grapple with problems. There are tens of thousands of years in the future. Our responsibility is to do what we can, learn what we can, improve the solutions and pass them on. It is our responsibility to leave the men of the future a free hand. In the impetuous youth of humanity, we can make grave errors that can stunt our growth for a long time. This we will do if we say we have the answers now, so young and ignorant; if we suppress all discussion, all criticism, saying, "This is it, boys, man is saved!" and thus doom man for a long time to the chains of authority, confined to the limits of our present imagination. It has been done so many times before.

—Richard P. Feynman, The Pleasure of Finding Things Out, Cambridge, MA: Helix Books, Perseus Publishing, 1999.

All at once it became vividly clear to Adam. He turned to the sated Eve—she was surrounded by apple cores—and wiped the apple juice from his chin with the back of his naked hand and remarked, "You know my dear, we are living in a time of transition!"

—Anonymous.

CHAPTER

20

Contents

Issue Resolution

20.1 Proposal

Indeed, we live in a time of transition as we observe the quality of our quality practitioners inexorably decline into mediocrity [1]. The confusions of the 1994 revision have been magnified by the complexity and redundancy of the 2000 version. The practitioners who found it difficult to grasp the concept that quotes were an integral part of contract review, that statistical techniques applied to corrective and preventive action data analysis, and that returned goods (that belonged to the customer) were part of customer supplied product will find it even more difficult to grasp the 2000 version concepts of organizational process structures. I have already seen clients roll their eyes in disbelief when auditors cannot grasp their elegant solutions to flowed-down quality objectives. Additionally, paraphrased manuals remain common; hours of ISO 9001:2000 seminars result in clients who still have no idea about how to design their QMS; and debates on continuous versus continual improvement waste valuable training time.

The 2000 version changes are so significant that sector-specific-standards based on the Standard require global revisions to maintain harmony with the Standard. In addition, because of these significant changes, most 1994 certified organizations have seriously delayed their upgrade activity. This delay will most probably lead to a mad rush in 2003 to bring thousands of sites up to date by the mandatory December 15 of that year. Even at this late date, I have only upgraded 10 clients to ISO 9001:2000, and I complete about 70 third-party audits a year.

We have found that the following changes have caused the most confusion and most difficulty in implementation listed (in no particular order):

- The numbering system from 20 elements to 5 sections into which are integrated the old 20 elements;
- The requirement for a process-oriented QMS that has a major impact on the way top management must view the enterprise from a TQM perspective;
- The requirement of top management to continually improve the QMS effectiveness, especially its processes;
- The need to clearly define and document measurable quality objectives and to have those objectives flow down through the organization;
- The contradiction between the need for only six procedures and the requirement that documents are needed to ensure the effective planning, operation, and control of its processes, in conjunction with the note that procedures can be documented or not;
- Failure to clarify what a process document is and the definition that a procedure is a specified way to carry out an activity or a process;
- The broad-ranging title of 4.2.4: Control of Records, yet a narrow range of records defined in the Standard;
- The requirement to enhance customer satisfaction;
- The requirement that the management representative promote an awareness of customer requirements throughout the organization;
- The need for top management to assess opportunities for improvement and the need for QMS changes during management reviews;
- The need to evaluate the effectiveness of training;
- The requirements to maintain and manage organizational infrastructure and work environments;
- The stress on customer feedback and effective customer complaint resolution;
- The extensive use of quantitative methods to evaluate continual improvement;
- An annoying degree of redundancy.

We, of course, hope that our attempt to define an effective and affective set of QMS design rules will help in some measure to provide the tools needed to address this set of challenging requirements. Our stress has been on the overall QMS design structure with multiple suggestions on how to create and optimize a QMS that directly conforms to the Standard's requirements. In particular, the use of quality policy statements in response to each SHALL will serve as a beacon to effectively highlight each of these requirements and lead to their resolution.

To successfully apply this structural beacon, we have demonstrated that the root causes of the observed QMS deficiencies in structure exist primarily in the tendency for authors to do the following:

- Lack clarity in their overall QMS structural design;
- Perform inadequate research into the reality of their system's performance;
- Place too little time in process-document creation;
- Place quality policy statements in lower level documents instead of in the chosen stand-alone manual;
- Paraphrase the Standard and leave out key prescriptive details for decision makers;
- Bypass SHALLS because of an incomplete analysis of the requirements;
- Use an integrated policy and procedure manual that does not fully respond to the Standard's requirements (SHALLS);
- Not stress the importance of tier-to-tier linkage;
- Maintain redundant procedures in several documentation tiers;
- Have no reliable source of interpretation to turn to.

This tendency is a result of an industry-wide disagreement by ISO 9000 practitioners on the purpose and structure of the manual. The result is confusion over what constitutes an effectively written document and what are the specific textual tools that we have to create the manual. Such practices are counterproductive because they invariably produce redundancy, omission, and noncompliance with the Standard.

To ameliorate this situation, we have attempted to place QMS documentation design and implementation on a scientific foundation. We have proposed a number of design rules that we believe produce compliant quality manuals—and, as a result, compliant quality management systems. Such systems integrate business strategy with quality management and thereby form the organization's total QMS strategic enterprise position.

Leadership is the ability to extend what is known while, concurrently, inspiring others to support you in your venture. Leadership is dynamic and self fulfilling. That is why you can achieve the same goals through widely different strategies. A dedicated group will meet their goals regardless of the leader's specific directives. A dedicated group will succeed even when the leader is less than adequate to the task. The leader cannot destroy the underlying professionalism of an individual. Many a leader has thought that they had been the motivator because the group succeeded, not realizing that the group succeeded in spite of its leader's incompetence.

As this unique human phenomenon is true in all human activity, it holds true in the creation of an effective QMS. Accordingly, our design rules may be used in any order that you see fit. More rules that are appropriate to your specific enterprise needs can be added. What is key, however, is that our design rules be used somewhere in the process as a foundation and context for the QMS. The rules will create self consistency and diminish redundancy as well as promulgate clarity and vigor throughout the entire creative process.

A summary of the design rules and specifics of such techniques is addressed in Chapter 21. In this chapter, we wish to examine the anticipated benefits of a QMS that uses the proposed design rules.

20.2 Benefits

The benefits to be gained from a QMS that is fully compliant with the Standard and integrates business strategy with quality management is summarized in Table 20.1. We have assumed that the fully responsive techniques discussed in this book have been chosen to create the QMS. Specifically, Table 20.1 considers the benefits to be gained for three functional categories:

- *Type I—readers:* for the readership, the benefits extend from improved communication to improved strategic and tactical decisions. Most importantly, all members of the value chain are included in this group.
- *Type II—organizational objectives:* with regard to organizational strategy, the impact is exceptional when we analyze the organization's ability to set, pursue, and communicate quality objectives. This enhanced communication begins with the posted organization's quality policy and propagates throughout the organization via the publication of progress reports on measured performance. Although measured performance was a

Table 20.1
Benefits of the Unified QMS (Fully Compliant with the Standard and Integrates Business Strategy with Quality Management)

Type of Reader or Function Type I—Readers	Benefits
Site manager	Significantly improved communication at all levels; opportunity to modify processes based on a more complete perspective
Executive staff	Obviously strong correlation between the completeness of the manual and the overall knowledge of the executive staff with regard to business policy
Customer	Dramatic improvement in communication and acceptance for more demanding contracts
Third-party registrars and assessors	Exceptional clarity leads to a far more effective assessment at a greater depth into the organization
Subsuppliers	Significantly improved grasp of your objectives and how to respond to them
All decision makers	The availability of clear and concise information significantly improves the decision-making process
Type of Reader or Function Type II—Organizational Objectives	**Benefits**
Response to organizational objectives	Exceptional response at all levels of the organization in the measurement and publication of enterprise metrics
	A powerful framework within which to establish quantitative quality objectives throughout the enterprise and to categorize them in terms of metrics and goals/targets
	A signal to all employees that the main purpose of the ISO 9000 certification is to improve the effectiveness of the operation, not just achieve certification
Type of Reader or Function Type III—The QMS	**Benefits**
Tier II documentation	Very strong influence on the completeness and effectiveness of hub documents and knowledge of business processes
Tiers III and IV documentation	Appears to have a minor effect. We have observed exceptional tier III performance with incomplete manuals

requirement in the 1994 version, it is more solidly and formally addressed in the 2000 version, especially with the focus on quantitative analysis.

- *Type III—the QMS:* with respect to the QMS, we observe a strong impact on the clarity and completeness of tier II documents. The tier II documents have been extended over the 1994 version to more clearly address the need for process documents that can be in the form of SOPs. However, the focus on processes opens up the doorway to a TQM evaluation of all of the organization's core competencies.

However, the impact on tier III documents as compared to tier II documentation is not that strong. This comes as no surprise because every organization, no matter how new or small, must work from some basic documentation and formatting. It is not unusual to find excellent work instructions and significantly incomplete manuals in the same organization.

Endnote

[1] Unfortunately, justifiable complaints against the ISO 9000 schema can be found throughout ISO 9000 publications (e.g., "ISO Warns Third-Party Industry to Police Its Self," *Environmental Management Report*, QSU Publishing Company, December 2001, p. 10. A more biting indictment is by Fahrlander, H. W. (Hank), Jr., "Don't Blame ISO 9000 for Poor Quality," *Quality Systems Update*, QSU Publishing Company, December 2001, p. 13.

CHAPTER

21

Contents

QMS Documentation and Implementation Design Rules

21.1 Design Rule Tables

We have now treated each element of QMS design in some detail and are in a position to summarize the set of design rules—that if applied by authors to the QMS creative process—should resolve all of the issues under consideration (see Table 21.1). The table is arranged in terms of the applied design rule and resultant benefits of the approach.

Summary of Global Mandatory Requirements As a memory jogger and very handy check-off list, the following global documentation has been shown to be mandatory (although there are requirements that are partially discretionary):

- A QMS;
- A quality manual (tier I);
- Documented quality policy (tier I);
- Documented quality objectives (tier I);
- Identification of processes (tier I);
- Sequence and interaction of the processes (tier I);
- Management reviews (tier I);
- Process plans (tier II);

Table 21.1
ISO 9001:2000 QMS Design Rules

QMS Design Rules	Benefits of Approach
Integrate business strategy with quality management—by means of a total business and quality policy format	Forms the basis of a well-informed organization Supports the organization's information technology imperatives Caters to decision making
Cleary define all of the organization's core competencies in terms of processes (e.g., marketing and sales, engineering, manufacturing, quality assurance, finance, and customer service)	Will be the framework for the process-oriented QMS design
Comply exactly with the Standard's requirements Respond positively with a quality policy statement to each SHALL, and adhere to the spirit of the ISO 9001:2000 requirements and guidelines	Ensures compliance with the Standard Enhances the inherent continuous/continual improvement cycle Enhances the possibility of payback
Utilize stewardship management with cross-functional teams Use experts to write sections	Ensures that top management is committed to the complete documentation, implementation, and demonstration of effectiveness of the program Partially satisfies the affective part of QMS design
Analyze all processes of the organization prior to the specific design decision Use flow charting methods if possible—otherwise tables and charts	Greatly enhances the development of the quality manual and provide an outstanding base for the hub documents
Formally declare the specific sequence pattern for the QMS layout (e.g., direct sequence with ISO 9001:2000 elements) and intensively train team members in the approach Select from four possible sequences: direct ISO 9001:2000, Shewhart cycle, operational cycle, and another standard's	Team members will work towards clear linkage to lower level documents Team members will concentrate on operational flow and the continuous/continual improvement cycle
Formally declare the specific manual configuration (e.g., stand-alone), and intensively train team members in the approach Select from either stand-alone or integrated configurations Be consistent with the placement of quality policy statements	

Table 21.1 (continued)

QMS Design Rules	Benefits of Approach
Evolve online in the shortest possible time Put level I and II documents online as early as possible Appeal directly to the customer/client's perspective	Leads to significant gains in document control and revision control Lowers distribution and maintenance costs Ensures clarity for all readers Caters to the new customer
Clearly place all quality policy statements in the quality manual Avoid redundant statements in lower level documents State the quality policy statements once	Creates a fully compliant manual that is clear and precise with regard to the organization's rules, methods, and business strategies
Include sufficient detail in the quality policy statements Allow the reader to understand how the organization actually works	Provides all readers, especially decision makers, with worthwhile pertinent organizational information Provides a highly effective document for the new customer
Provide user-friendly navigation tools A four-tier structure Hub documents	Significantly increases the probability of effective implementation by all employees Tends to minimize the number of documents Clarifies linkage
Avoid paraphrasing	Removes the trivialization of the Standard that is anathema to the quest for continuous/continual improvement where completeness and clarity are mandatory
Use effective styles such as the following: Simple declarative sentences in the active voice and present tense Avoid redundancy Clear labels Useful TOC Minimize jargon Stress clarity Define terms Effectively link Avoid "ing" usage Use bullets Don't rely on the spell checker Use graphics	Enhances information flow Increases rate of understanding Increases training retention time Increases rate to find documents
Use the same design rules for sector-specific requirements	The exact same benefits apply

Table 21.1 (continued)

QMS Design Rules	Benefits of Approach
Use as many words and charts as is needed to produce the organizational image that is desired	A little long May be too strong— But it ain't wrong Terse is worst!
Utility—documents should be as follows:	Enhanced communication
Worth reading	A common training vocabulary
Contain industry familiar phrases	Propagates the quality policy directive to all employees
Relatively easy to obtain	
Online considerations:	Employees tend to get confused immediately when asked to find a specific file unless they can quickly enter a master list of documents
Use user-friendly entrance icons to the QMS documentation structure (e.g., ISO 9000 QMS link)	Lowers user anxiety through familiarity and is a quick entrance to a master documentation list
Keep separate directories for each department, core competency, or process document	A very easy and effective technique for QMS documentation navigation
Use the manual cover page as a hub document with hyperlinks to lower tier documents	Removes user anxiety—a simple cheat sheet Especially important when flow charts are used Topics such as electronic signature control are to be included
Provide a controlled desk reference chart on how to enter and navigate the QMS documentation structure	
Make certain that all terminals are provided with the proper software applications and that all terms are compatible with the network	
Be sure to create a procedure that covers security/password control, backup, and configuration management details for the QMS online system; remember to validate software when appropriate	

- Monitoring, measuring, analysis, and improvement plans (tier II);
- Six documented procedures (tier II);
- Work instructions as applicable (tier III);
- Other documents needed to ensure the effective planning, operation, and control of its processes (all tiers);
- Records as required (exist at all tier levels)—to indicate objective evidence of effective operation;

- Documents that describe product characteristics (tier III);
- Method of linkage between tiers;
- Declaration of the ISO management representative;
- Description of the organization to be certified;
- Specific description of responsibility and authority of at least the top management;
- Inter and intra organizational interfaces;
- Demonstration of the effective implementation of the system;
- The use of sensible levels of documentation;
- The effective management of customer complaints;
- Declaration of factored items (if applicable);
- Master list of current Standards and codes.

21.2 Closing Invitation to the Case Studies

This completes our ISO 9001:2000 QMS Documentation Design and Implementation design rules. We believe that the use of the proposed rules will result in a more user-friendly, effective, and affective QMS.

We have approached the subject of QMS design as a scientific exercise—although there is a very significant subjective aspect to the paradigm. In this sense, we are reminded of the words of Thomas S. Kuhn [1]: "Probably the single most prevalent claim advanced by the proponents of a new paradigm is that they can solve the problems that have led the old one to a crisis. When it can legitimately be made, this claim is often the most effective one possible."

We trust that we have presented a legitimate discourse. In fact, it is now time to apply the ISO 9001:2000 QMS design rules to the ambitious and fast growing Growth Corporation, now a wholly owned subsidiary of the Stable Corporation, which has decided to upgrade from ISO 9001:1994 to ISO 9001:2000 using the 2000 format. The Growth Corporation was certified to ISO 9001:1994 3 years ago, and the time has come for its recertification. The timing is such that they might as well upgrade to ISO 9001:2000 as part of the final surveillance on their present contract and save money on one certificate change instead of waiting until 2003 to upgrade. Let's see how they manage their upgrade mechanics based on the cut-and-paste and fill-in technique [2].

In addition, let's also see how Growth helped a fellow company create a QMS:2000 on its first certification experience based on the Growth Corporation's previous two certification experiences.

Please join us now in the case studies. Hope you enjoy the presentations as much as I did their creation. Besides, you might find someone who sounds like you!

Endnotes

[1] Kuhn, Thomas S., *The Structure of Scientific Revolutions, Foundations of the Unity of Science,* Vol. II, No. 2, University of Chicago, 1970, p. 153. The continued interest in Kuhn's paradigm shift theory has been enhanced with the addition of the following publications as noted by Wilson, Kennith G., *Physics Today,* March 2001, p. 53: Kuhn, Thomas S., *The Road Since Structure: Philosophical Essays, 1970–1993,* with an autobiographical interview edited by James Conant and John Haugeland, Chicago: University of Chicago Press, 2000; and Fuller, Steve, and Thomas Kuhn, *A Philosophical History of Our Times,* Chicago: University of Chicago Press, 2000.

[2] The certification of the Growth Corporation to ISO 9001:1994 was detailed in: Schlickman, Jay J., *ISO 9000 Quality Management System Design, Optimal Design Rules for Documentation, Implementation, and System Effectiveness,* Milwaukee, WI: ASQ Press, 1998.

PART

VII

Two Case Studies

The secret of the master warrior is knowing when to fight, just as the secret of the artist is knowing when to perform. Knowledge of technical matters and methods is fundamental, but not sufficient to guarantee success; in any art of science of performance and action, direct perception of the potential of the moment is crucial to execution of a master stroke.

—Thomas Cleary in The Lost Art of War by Sun Tzu II, San Francisco: Harper, 1996, p. 52.

Whatever I find myself doing, I become aware that I must make a choice. I must make a choice or find the choice made for me. I must choose from whatever alternatives my experiences have stored up and from whatever alternatives my emotions made available to me. I must try to calculate the risks involved, and manage my fears while calculating.

—John C. Glidewell, Choice Points, Cambridge, MA: The MIT Press, 1972, p. 3.

CHAPTER

22

Contents

Case Study #1: The Growth Corporation Upgrades to ISO 9001:2000

22.1 Choice Point

22.1.1 Author's Introduction

The result of the effort to create an effective Growth Corporation Quality Management System (GCQMS) Manual is the gist of this first case study. The pages in this book, from the GCQMS cover page in Section 22.3 until the end of the case study in Section 22.8, form a contiguous ISO 9001:2000 manual that conforms in detail with the Standard.

The screened type represents the text from the 1994 manual that was described in my previous ISO 9001:1994 application-oriented book.

The black type represents the additional responses needed to bring the old manual into conformance with the Standard. As you scan the pages you can readily discern that Section 22.4 contains the greatest amount of additional material. Because there are over 420,000 1994 manuals already written, this case study should give you an excellent idea of what your upgraded manual will look like. For those who are creating their first manual, it is a practical example of what your manual should look like if you apply the design rules described in this book.

Interestingly enough, the design rules that had been used to create an effective QMS:1994 have been essentially invariant under the QMS:2000 transformation. This is not surprising because the design rules are generic to the extent that they are applicable far and beyond ISO 9000 documentation and implementation. For example, the concept of shall analysis is applicable to any proposal/quote written against a given set of requirements typical of a Department of Defense or NASA request for proposal (RFP) or request for quote (RFQ). Moreover, even if you do not have a copy of my ISO 9001:1994–oriented book, you can readily determine how to create the upgraded QMS by studying this book and also analyzing the second case study, which follows Growth's first certification process in detail.

Now, let us look in on Fran and Mike as they make the decision to go ISO 2000.

22.1.2 An Upgrade Decision

Growth's fifth maintenance surveillance came to a highly successful conclusion—zero nonconformances. In addition, several interesting and valuable opportunities for improvement were offered by the lead assessor, who was now on his way to the next audit.

Fran, the president and CEO of Growth, Inc., remained in the same conference room where the closing meeting had taken place, along with Mike, the vice president of quality assurance. Both were relaxed and satisfied with the surveillance's results and were sipping down some diet cola and munching on the large sugar cookies that were a staple at Growth. The ambiguity between lowered calories in the cola versus a gigantic load of calories in the cookies never seemed to be an issue at Growth. People needed that sugar!

Fran began the dialogue. "Mike, it seems impossible. We're only 6 months away from the recertification audit. Where did those 3 years go?"

Mike smiled warmly, "Well we did a lot in that time. We got certified, we got raised to a wholly owned subsidiary—as we planned—we've landed some really fine OEM contracts, we've pulled off that cost-reduction program that was really key to our profitability, and we both have 'president' in our titles. Not bad, I'd say."

Fran agreed completely with Mike's quick summary of their progress. The past was fine, but where to go from here was her main concern. "Tell me, Mike," Fran questioned, "I've heard a great deal about the new ISO version—especially from you. Shouldn't we upgrade to ISO 2000 about now?"

Mike pondered the question a bit, then replied, "Well, we have some time before December, 2003, when it becomes mandatory. However, in six months, when we recertify, we get a new certificate. I've been talking to Sam about what needs to be done, and he feels that we could pull it off in time to have it added to our recertification audit. He feels that a technique called 'cut--and-paste' is the way to go. I've already taken a two-day class on the ISO update, and what he says makes sense to me."

Fran wasn't surprised that Mike had already prepared himself for the upgrade effort by contacting their consultant and taking courses, but she wasn't sure that the staff was ready so relatively soon after the last certification. She replied, "Mike, do me a favor, prepare a 30-minute briefing for the staff for Wednesday's ISO management review meeting and let's see how everyone feels about this."

Mike immediately assembled his notes and prepared for the review. He felt it was extremely important that Growth remain up to date in the quality arena, and the upgrade would be a effective marketing device—especially because so many companies were dragging their heels in this respect.

22.1.3 The Staff Meets

The Growth executive management team (GEMT) monthly management review meeting began on schedule and covered the usual items related to progress against quality objectives and the corrective and preventive action program. This took about 45 minutes. Fran then introduced the ISO 9001:2000 upgrade possibility and asked Mike to give his presentation. As soon as Mike started, everyone groaned, "Oh, not again!" It was not exactly good natured.

Mike explained that they would have to do it anyway in about a year, and why not save a few bucks on the certificates by doing it at the recertification audit. He pointed out that Sam had estimated the effort to take only four months and to not use up more than about 160 actual hours for the GEMT staff, or about 23 hours per person over the 4 months. This was because the company was already TQM-oriented, and the toughest section on quality objectives, metrics, and targets was already in place.

Most of this time would be spent on a rewrite of the quality manual. The ripple effect to the lower tier documents would be minimal and should require no more than another 160 hours of employee time.

At the moment, Mike couldn't see why it would take any more of an effort because the group already thought in terms of core competencies; the audit program was already process oriented; the corrective and preventive action program was already geared towards customer satisfaction; and management review meetings already included all of the prescriptive requirements of the new revision, and then some. Growth had also been blessed with several very competent third-party assessors, who had consistently fine tuned their QMS, and Growth had responded vigorously and successfully to half a dozen customer audits during the past three years. With respect to the 2000 version, the present QMS was hot cherry pie in the pan, cooking in a smoking kitchen!

The staff trusted Mike and had little comment, except that there wasn't a lot of time between now and the recertification audit, so they had better get started. Each staff member agreed to be process champions, as they had been before, and to select subprocess champions as required. Mike thanked the group, contacted Sam, and launched the cut-and-paste upgrade project.

22.1.4 The Upgrade Assessment

The upgrade effort proved to be highly successful. The assessors first examined the quality manual offsite. They had very few comments to make about the obvious conformance of the manual with the Standard. As a result, a preassessment was deemed unnecessary, and Mike decided on just going for the upgrade assessment about 30 days after he had received the offsite document review report from the lead assessor.

The upgrade certification assessment was carried out the day after the sixth surveillance was completed and went extremely well. The GCQMS was already in strict conformance with the Standard due to perceptive surveillance audits by the lead assessor and an intensive internal audit program. The three minor findings were really tune-up type observations (e.g., it was felt that the method of amendment to requests for proposals was not clearly stated). Mike took care of the nonconformance by adding a sentence to the manual (i.e., "The new sales orders are also required for RFP amendments"). The other two nonconformances had to do with quotes that are stored electronically but not clearly addressed and manufacturing capacity was not clearly addressed as a part of the quote process. Mike readily fixed those up, too. There were also a number of opportunities for improvement suggested by the assessor, and Mike promised to take them into consideration. Mike felt that each one had merit and included all of them into the GCQMS by the first surveillance assessment 6 months later.

At the closing meeting, the assessor was delighted to inform the company that the recommendation would be for upgrade to ISO 9001:2000. Growth could tell the world that they had been recommended for approval to the Standard. Approval by the registrar's main office would follow within 30 days. This was the assessor's tenth upgrade, and he hadn't lost one yet. Surprisingly, the executive team cheered out loud and then decided to hold a beer and pizza party that night. They weren't jaded after all.

We will now see how the team did the job. We begin the dynamics of the upgrade with some application notes.

22.2 Application Notes to the Upgraded Quality Manual

The key quality manual design rules adhered to are as follows:

- It is a stand-alone document.
- It uses the Standard's numbering system (see Sections 4–8).
- It responds to each SHALL in the Standard.
- It was created using the cut-and-paste method.
- It includes sections that were created by subject matter experts.
- It clearly states responsibility in each section.
- It ensures that sections/subsections refer to the appropriate hub document.
- It did not discuss proprietary information.
- It contains a description of Growth's business and its vision, mission, and quality policy statement to capture the concept of an integrated business and quality strategic declaration.

The following general documentation design rules were followed:

- The manual is based on a four-tier system.
- Beginning with Section 4, the screened type represents the previous 1994 quality manual text (i.e., screened type implies old text).
- Black type represents ISO 9001:2000 additions to bring the 1994 manual in conformance with the 2000 requirements. A quick scan of each section shows clearly the scope of upgrade required by comparing the frequency of the black type.
- Hub documentation system used.

The following sections begin with the cover page and table of contents for Growth's upgraded ISO 9001:2000 quality manual, followed by Growth's response to the requirements of Sections 4 through 8 of the Standard. The numbering system is in a 1:1 correlation with the Standard. The set of Sections 4 through 8 represents a manual that is fully compliant with the Standard and uses the design rules defined in this book to achieve conformance.

22.3 The Upgraded ISO 9001:2000 Quality Manual: Cover Page and Table of Contents

Growth Corporation Quality Manual

St. Louis, MO

GCQMS Navigation Linkage

First-Time Users Begin with Tier I, Quality Manual Table of Contents: TOC TOC

Expert Users If you wish to directly view the other informational tiers of the QMS first, click the hyperlinked titles to move directly to the appropriate master lists to acquire the necessary documentation:

- Tier I—Quality Manual Table of Contents TOC
- Tier II—HUB Documents ..\\ISO DOC T2\\Master List.doc
- Tier III—Work Instructions ..\\ISO DOC T3\\Master List.doc
- Tier IV—Forms Master List
  ..\\ISO DOC T4\\Engineering Forms\\Master List.doc
  ..\\ISO DOC T4\\General Forms\\RECMSTER1.doc
- Records Master List
  ..\\Records\\RECORDS MASTER LIST.doc
  ..\\Records\\DeptMinutes\\MASTER LISTMinutes.doc
- Corrective Action Master Lists ..\\NCR's\\NCR Master List.doc
- Preventive Action Master Lists ..\\NCR's\\NCR Master List.doc
- Nonconforming Material Master Lists ..\\NCR's\\NCR Master List.doc
- Supplier CAR Master Lists ..\\NCR's\\NCR Master List.doc
- Reports Master Lists
  ..\\Reports\\XBus Mgt Master List.doc
  ..\\Reports\\YPM Master List.doc
  ..\\Reports\\ZSales Mgt Master List.doc
  ..\\Reports\\ZSales Mgt Master List.doc
  ..\\Reports\\ZSales Mgt Master List.doc
  ..\\Reports\\ZSales Mgt Master List.doc

..\\Reports\\ZSales Mgt Master List.doc
..\\Reports\\ZSales Mgt Master List.doc

- Audits Master List ..\\Audits\\Audit Master List.doc
- Internal Communications Master List ..\\Internal Communications\\Int Comm MASTER LIST.doc
- Quality Objectives—Metrics and Charts ..\\Internal Communications\\Int Comm MASTER LIST.doc

This is a controlled online document. Neither electronic nor hard copies of such documents may be made or distributed to any person or persons, companies, or organizations, without strict adherence to both Growth's *document control procedures and disposition of proprietary information.* Growth *reserves the right to change, modify, and/or withdraw this document, without notice, other than under those conditions specified by the ISO 9000 Registrar.*

Approval: *Fran Dewolf*	First Release Dated: January 1, 2002
President and CEO	Revision: 03
The Growth Corporation	

GCQMS Quality Manual
Table of Contents

7.4	Purchasing	—	Title only
7.4.1	Purchasing Process	—	4.6.2
7.4.2	Purchasing Information	—	4.6.3
7.4.3	Verification of Purchased Product	—	4.6.4, 4.10.2
7.5	Production and Service Provision	—	Title only
7.5.1	Control of Production and Servicing Provision	—	4.9, 4.15.6, 4.19
7.5.2	Validation of Processes for Production and Service Provision	—	4.9
7.5.3	Identification and Traceability	—	4.8, 4.10.5, 4.12
7.5.4	Customer Property	—	4.7
7.5.5	Preservation of Product	—	4.15.2, 4.15.3, 4.15.4, 4.15.5,
7.6	Control of Monitoring and Measuring Devices	—	4.11.1, 4.11.2
8.0	Measurement, Analysis, and Improvement	—	Title only
8.1	General	—	4.10.1, 4.20.1, 4.20.2
8.2	Monitoring and Measurement	—	Title only
8.2.1	Customer Satisfaction	—	New
8.2.2	Internal Audit	—	4.17
8.2.3	Monitoring and Measurement of Processes	—	4.17, 4.20.1, 4.20.2
8.2.4	Monitoring and Measurement of Product	—	4.10.2, 4.10.3, 4.10.4, 4.10.5, 4.20.1, 4.20.2
8.3	Control of Nonconforming Product	—	4.13.1, 4.13.2
8.4	Analysis of Data	—	4.20.1, 4.20.2
8.5	Improvement	—	4.1.3, 4.14.1, 4.14.2, 4.14.3
8.5.1	Continual Improvement	—	4.1.3
8.5.2	Corrective Action	—	4.14.1, 4.14.2
8.5.3	Preventive Action	—	4.14.1, 4.14.3

22.4 Quality Management System (QMS)

4.1: General Requirements

QMS Responsibility The Growth Corporation (Growth) quality management system (GCQMS) is defined in its quality manual (manual). The Growth executive management team (GEMT), which consists of the president and direct reporting staff, is responsible for the extent and content of the manual. The manual is maintained and kept current by the director of quality assurance, who also serves as the ISO 9000 management representative.

It is the direct responsibility of the GEMT to ensure that there is a continual improvement in the effectiveness of the GCQMS through constant oversight of its processes. The requirements of the ISO 9001:2000 Standard (Standard) are used to establish the framework within which these processes interact.

Eight Quality Management Principles The GEMT applies the eight quality management principles defined by the Standard as a means to ensure continual performance improvement in the GCQMS. Specific activities that relate to each principle include the following:

- *Customer focus*—an extensive customer service organization to complement and supplement marketing and sales continuous analysis of customer response and complaints based on periodic customer satisfaction surveys.
- *Leadership*—an annual business plan prepared by GEMT that includes quantitative quality objectives, metrics, and targets for all managers and supervisors.
- *Involvement of people*—the use of cross-functional teams in the setting of quality objectives and the solution of process and product nonconformances.
- *Process approach*—the establishment and analysis of core competencies for each company function.
- *System approach to management*—the integration of business objectives with quality objectives so that the company processes represent a total quality approach to continual improvement.
- *Continual improvement*—the goal of all employees based on quality objectives and intense training in quality management system concepts and implementation.
- *Factual approach to decision making*—decisions primarily based on the analysis of quality objective progress against targets. The GEMT reserves the right to make decisions based on less statistical information when appropriate. Decision making at Growth is a holistic process that incorporates all available internal and external data.
- *Mutually beneficial supplier relationships*—a vigorous supplier partnership program that provides suppliers with periodic evaluations of their performance and the necessary support to aid in nonconformance resolution.

Exclusion Statement The GCQMS is fully responsive to all requirements of the Standard. There are no exclusions.

Growth's Core Competencies Growth has structured its company organizational processes in terms of core competencies. The core competencies are as follows (with main process champions noted):

- Executive (GEMT)—president and CEO;
- Finance and administration—controller;
- MIS management—MIS manager;
- Sales and marketing—vice president of sales and marketing;
- Design engineering—vice president of design engineering;
- Quality assurance—vice president of quality assurance;
- Manufacturing—vice president of manufacturing.

Processes Defined The full set of Growth business processes consists of the core competencies (main processes) and their associated subprocesses. The structure of the full set of processes is illustrated below in Figure 22.1. The set of process documents shown in this figure are available online in a secure intranet network. The document control subprocess provides protocols for hard copy, as necessary.

Process Sequence and Interaction Defined Figure 22.2 describes graphically how the main processes are designed to produce customer-desired products and how customer feedback is obtained and analyzed within the GCQMS structure.

Figure 22.2 illustrates the following processes:

- The development of product specifications is a joint effort of marketing and sales with design engineering. Design engineering interacts with the customer only when engineering is part of a design engineer and product manager team.
- Design engineering hands off production packages to manufacturing by means of cross-functional team relationships and the engineering change order process. The packages contain the required test fixtures and instrumentation needed to produce customer-specified product characteristics. The MIS department manages the intranet system that integrates all of the QMS operational functions, including those required by engineering design, quality assurance, manufacturing, and customer service.
- The team of design engineering, purchasing, and quality assurance evaluates and selects vendors. Subcontractors are chosen on an as-needed basis by the department heads.

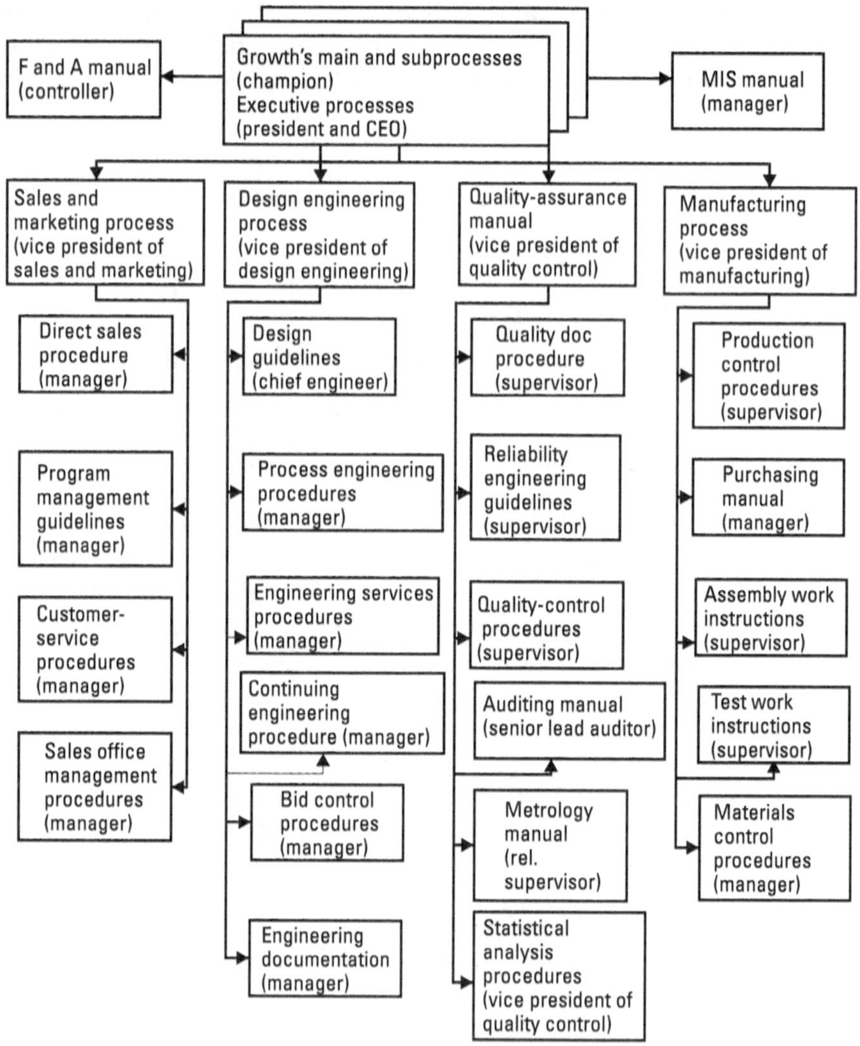

Figure 22.1 Growth's process management and quality management system.

- Manufacturing works with continuing engineering to optimize its processes as a means of shipping on time to desired performance levels. Quality assurance supports the calibration of measuring and monitoring devices in both manufacturing and engineering.
- After-sales activities and installations are managed by customer service under marketing and sales oversight. Customer service has a direct contact with the customer via returned goods and also orders spare parts from manufacturing and sells them directly to the customer. Customer service data is provided on a daily basis to quality assurance.

Figure 22.2 Growth's sequence and interaction of its main processes (refer to Figure 22.1 for subprocess interaction).

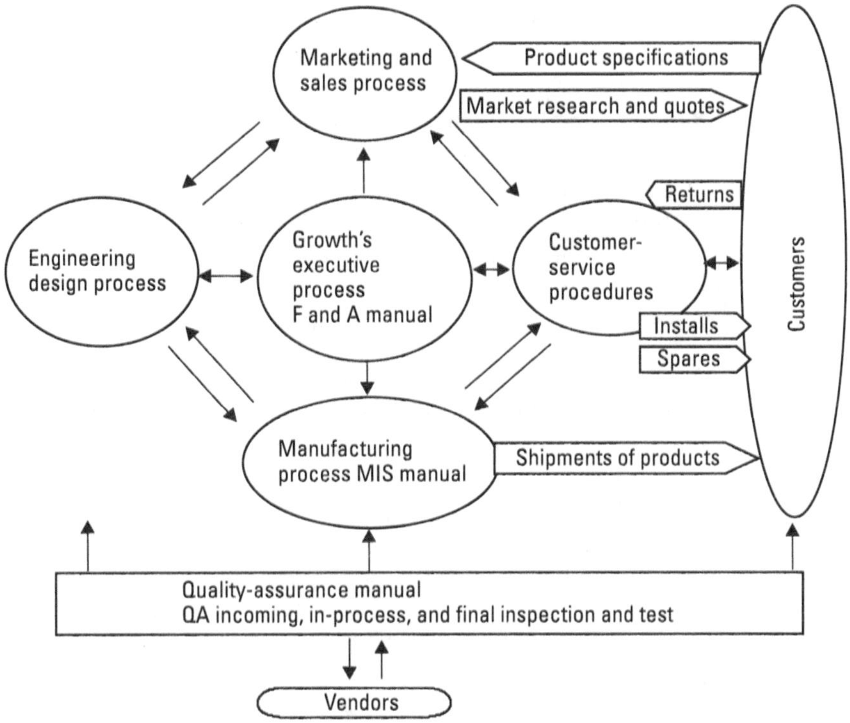

- Quality assurance works across the board to ensure the operational integrity of the processes and performs reliability studies, manages metrological activities, performs audits of the company and of vendors, and provides the cost-of-quality analysis for the company in concert with finance and administration. The department serves as part of the design engineering cross-functional team and performs incoming, in-process, and final inspection and testing.
- Finance and administration and MIS manage the online computer systems that are used by all departments.
- The GEMT uses a series of reviews to ensure the integrity and efficiency of the total management process. The reviews include the quarterly management review supplemented by monthly department reviews and weekly operational reviews. All of the reviews are documented and kept as records by either the ISO 9000 management representative or the appropriate local area manager.

Criteria and Methods The GEMT formulates quality objectives, their metrics, and targets as part of the annual growth business plan. This is a way to establish the set of criteria and methods to be used to effectively manage Growth's main processes and subprocesses. 🗁GBP.Doc

Resource and Information Availability The Growth business plan also establishes the capital and personnel resources required to maintain an effective QMS and creates the framework for companywide information sources that include management reviews at all company levels; a monthly company newsletter; and quarterly company meetings by the GEMT for all employees.

Process Monitoring, Measuring, and Analysis

An intensive computer-aided program is used to analyze data obtained through the monitoring and measurement of key process parameters. The manager of statistical analysis is responsible for this function, which includes the corrective and preventive action process located in the quality-assurance manual. 🕮QAManual.Doc

Continual Process Improvement The GEMT uses a cross-functional action team structure to analyze and resolve process improvement issues. Oversight is accomplished both through specific action team reviews and quarterly GEMT reviews. The action teams complement the corrective and preventive action program. All action team activities are documented. 🗍TEAMS.Doc

Outsource Management Printed circuit board fabrication and painting are typical processes that Growth regularly outsources. In addition, circuit board layout and mechanical design are typically outsourced by engineering on an as-needed basis. Control of such outsourced processes is coordinated by the purchasing supervisor as part of Growth's supplier partnership program. All outsourced engineering projects are managed by the pertinent project engineer and regularly reviewed by the chief engineer. 🕮SPP.Doc

4.2: Documentation Requirements

4.2.1: General Requirements

The GCQMS includes the following:

- A controlled quality policy document that is posted as both an electronic file and on hard copy about the facility. Refer to Section 5.3 of this manual.
- A controlled set of quality objectives with metrics and targets based on the Growth Business Plan. Refer to Section 5.4.1 of this manual.
- The six procedures required by the Standard:

1. Control of documents in two procedures (i.e., QA document control and engineering document control). 🗁QADC.Doc and 🗁ENGDC.Doc
2. Control of records in the procedure (i.e., records control). 🗁Records.Doc
3. Internal audit in the procedure (i.e., auditing manual). 🕮Audits.Doc
4. Control of nonconformity in the quality-assurance manual. 🕮QAManual.Doc
5. Corrective action in the quality-assurance manual. 🕮QAManual.Doc
6. Preventive action in the quality-assurance manual. 🕮QAManual.Doc

Additional Documentation The GCQMS also includes a large number of other documents whose purpose is to ensure the overall effective planning, operation, and control of the QMS. The extent of the mandatory documentation and the supplemental documentation is explained next.

Life Cycle The QMS documentation is designed to impact the entire life cycle of Growth's hardware and software products. As a result, the product plans are based on the effective inclusion of all aspects of the product's life (i.e., from market share to after-sales service).

Four-Tier Structure The documentation is primarily online at the tier I and tier II level, and is a hybrid system otherwise (i.e., a mixture of electronic and hard-copy files). The system is illustrated in Figure 22.3. The online system includes an MRP system used by manufacturing and engineering.

Records As indicated in the pyramid, records can occur at any level and form their own particular documentation hierarchy. (Refer to Section 4.2.4 of this manual.)

4.2.2: Quality Manual

As illustrated in Figure 22.3, Growth's quality manual (manual) is the highest level document in the QMS. It defines Growth's quality policy statements for all five sections of the Standard and those portions of ANSI/ISO/ASQ Q9000-3-1997 that are applicable.

The president is responsible for the review and approval of the manual. As with all Growth documents, the manual is reviewed and updated either upon revision or during the quality audit process.

Figure 22.3 Growth's four-tier operational pyramid.

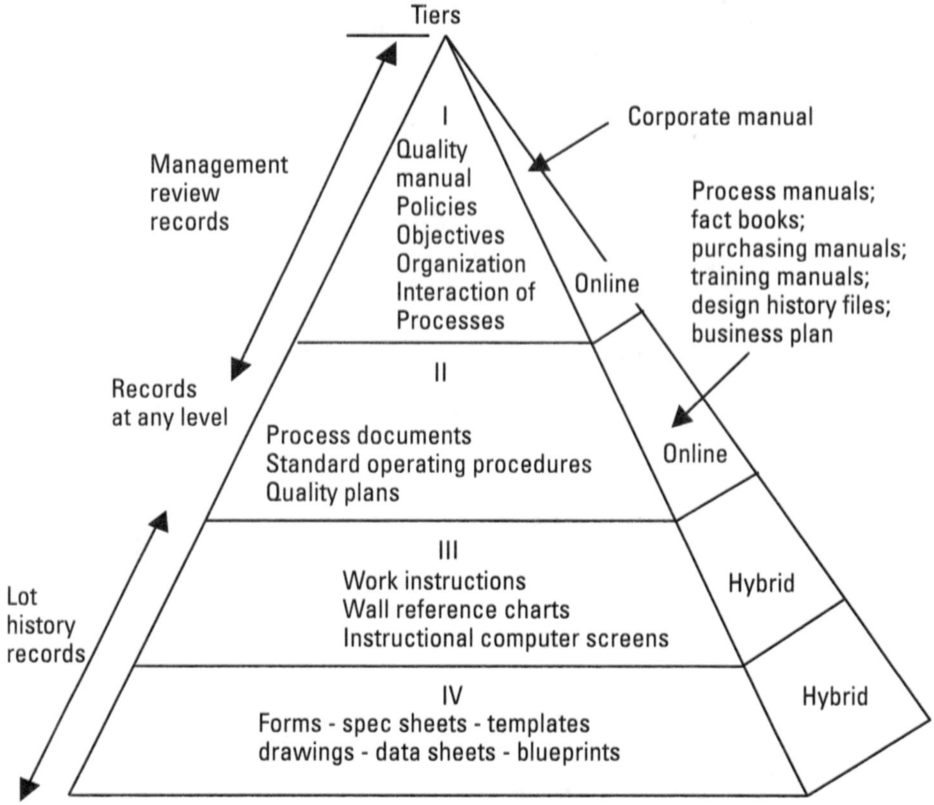

Scope The manual covers the design, manufacture, marketing, selling, and servicing of modular hardware and software products as a means to display and process commercial and industrial images on personal computers and workstations. As previously noted, there are no exclusions to the Standard's requirements.

Process interaction: Defined in the previous chart entitled, Growth's sequence and interaction of its main processes.

Linkage: Linkage from document to document within the QMS is by means of hyperlinks for the electronic files and references to hard-copy documents. Only the document title is used (i.e., electronic files are not numbered). In this manner, the reader is directed from the manual to the process documents and then to procedural and format documents, as appropriate. Hard-copy documents such as drawings and schematics are controlled numerically under engineering change order document control.

To expedite navigation, it is always advisable to begin with the manual and then go directly to the Appendix, entitled "Growth's Master List of Hub

Documents." The hub document can be likened to an airport hub. Once you reach the hub document, it leads the reader to the next levels of system information. As an aid to navigation, the manual's cover page contains key hyperlinks.

Procedures defined: It is important to note that Growth's "procedures" are in the form of either very high-level process documents or lower level procedures or work instructions. The high-level process documents are equivalent to standard operating procedures and are in the form of flow charts with a supplemental text document attached. The two documents form a single process document.

Implementation: The effective implementation of the QMS is ensured through a comprehensive management review defined in Section 5.6 of this manual. Most importantly, great care is taken to track all preventive actions achieved by Growth employees and to reward such activities commensurate with their contribution to the overall company's productivity.

Skill levels: All of Growth's documents are created to serve highly skilled and extensively trained employees. Moreover, Growth's employees are required to work for lengths of time without close supervision and to carry out multitasking work. As a result, the level of detail in the documents varies from engineering guidelines to detailed test protocols based on the specific operational and administrational tasks. Most importantly, the documentation is designed to support minimal supervision by being readily available yet unobtrusive.

4.2.3: Control of Documents

Procedure The description of Growth's approach to the control of all hardware and software documentation and data is contained in two documents (i.e., QA document control and engineering document control). QADC.Doc and ENGDC.Doc

Responsibility Control of documents is shared by the quality documentation supervisor and the engineering documentation manager. Both functions maintain master lists of documents. The master lists of documents are maintained to ensure the correct distribution of documents and that users have the most recently revised documents at their work sites. QCMLDOC.Doc; and ENGMLDOC.Doc

Control of Externally Received Documents Documents such as the ISO 9000 Standard, vendor documents, and customer specifications are controlled locally by the appropriate user. Such documents are used either in the design, manufacture, or preventive maintenance of product and associated instrumentation.

Document and Data Approval and Issue Control of the documents is maintained in the following ways:

- All released engineering documents are under engineering change order (ECO) control whether on hard copy or online. Program managers and project engineers have ready access to such documents via the engineering computer system (ECS). All policy, process, and procedural documents are controlled by means of either the document's name or by the code numbers and "red" control numerals, as necessary. Each process document is assigned a "champion" who is responsible for the review, approval (sign off), and release of the documents.
- Tier IV documents (e.g., forms contained in a forms master manual) are controlled by the quality documentation supervisor and distributed to the local managers upon request.
- Memos, reports, and similar documentation are controlled at the local-manager level and do not require a master list.
- Obsolete documents are removed from use at the local-manager level upon receipt of revised documents.
- Documents that are maintained for legal and informational purposes are marked accordingly on their containers and archived and maintained by the accounting department, or by the chief engineer, as appropriate. Department managers are authorized to determine that obsolete documents are to be retained.

Document and Data Changes Revision control is handled in the following manner:

- Changes at the policy, process, procedural, and forms level are made via the department change order (DCO).
- Changes to released hardware and software engineering documents are incorporated via the ECO.
- Document and data change orders for hardware and software products are reviewed and approved by the designees defined in the process and are usually the document owners or their designees.

- Previous revisions and pertinent background information are directly available as part of the DCO and ECO formats. Such formats also contain descriptive material for the nature of the change.
- Revisions to documents occur as part of the corrective and preventive action program, [e.g., audits, nonconformance material reports (NCMRs), corrective action reports (CARs), and preventive action reports (PARs)]. However, all documents that have not been revised for over two years are reviewed for currency by the document's champion.
- Online document legibility is inherent. Hard-copy documents are kept in steel case files and/or banker's boxes when stored. Internal audits ensure that document deterioration is minimized.

Online control is handled differently. The online documentation systems are secured via passwords and periodically backed up and stored electronically by the MIS manager. 🕮MISMANUAL.Doc

4.2.4: Control of Records

Procedure The records control procedure and its associated records master list detail the procedures for identifying, collecting, indexing, accessing, filing, storing, maintaining, and disposing of quality records. 🗁RECORDS.Doc and 🗍RECORDSML.Doc

Responsibility Maintenance of records is the responsibility of each local area manager. However, for the sake of continuity, the ISO 9000 management representative maintains the records master list that is an online summary of all core competency master lists.

Control Quality records are identified by either title or number. The records master list indicates the location of the records. They may be online or hard copy. The master list includes who is responsible for the record, its retention time, and its status (online, hard copy, obsolete, or obsolete but retained).

Subcontractor data, in the form of certificates of compliance or analysis, is included in the master list.

Filing The hard-copy records are filed in either commercial-grade steel case files or in corrugated containers such as banker's boxes. Online records are managed by the MIS manager. All files are kept on site. Records are kept within the easy reach of the users to facilitate usage.

Legibility The hard-copy quality records are typed for legibility whenever possible. Handwritten records are written with permanent black ink pens. Online records are inherently legible.

Conformance Growth uses its quality records as a key source of information for presenting the status of its internal audit and preventive and corrective action program to management for review. Most importantly, the records are a source of quantitative information used in trend analysis.

Process Control Many records are kept by the local area manager (e.g., process control records are kept in the individual customer job folders in production and include, when available, assembly drawings, bills of materials, fabrication drawings, visual aids, process control documents, and measurement documents).

Contractual Agreements In those cases when Growth enters into a contractual agreement, records are made available to the customer or its representative for evaluation for a limited period.

Disposition Because Growth's records consist of records from all of the company's core competencies, the disposal of records requires controller approval, and, in the case of company proprietary information, the approval of the president.

22.5 Management Responsibility

5.1: Management Commitment

Profile The Growth Corporation (Growth) was established in 1993 as an operating division of the Stable Corporation (Stable), a business enterprise founded in 1985, with corporate offices in Dallas, Texas. Growth has been charged by Stable's board of directors to develop and market hardware and software accelerator boards for data acquisition purposes across a very wide range of commercial and industrial applications. Due to successful achievements of specific productivity and profitability goals, Growth became a wholly owned subsidiary of Stable in 2000.

Growth's first series of microlayered board, box, and subsystem components were very well received by OEMs and this allowed the company to rapidly increase sales over the past seven years. Growth's ANSI/ISO/ASQC Q9001-1994 certification in 1998 was indicative of the division's desire to continue this impressive growth in concert with increased quality and with customer satisfaction foremost in mind. Today, Growth will take the next step to total quality management as it upgrades its quality management system to the 2000 version of the Standard.

Vision Growth's vision is to be the leading supplier of microlayered board, box, and subsystem technology for imaging systems.

Mission To achieve this goal, Growth will

- Work to continue to increase its partnership basis with customers to satisfy their technological needs;
- Continue to design products that meet the customer's explicit and implicit requirements;
- Manufacture products that are delivered against the customer's on-time requirements;
- Respond quickly and thoroughly to customer complaints and service requests;
- Maintain an effective ANSI/ISO/ASQC Q9001-2000 quality management system that also complies with the applicable clauses of the ANSI/ISO/ASQC Q9000-3-1997 computer software guidelines;
- Meet its financial goals in agreement with Stable's corporate requirements;
- Provide a responsive, rewarding work environment for its employees.

Business Objectives To satisfy our mission requirements, Growth will

- Periodically survey our customers to establish both satisfaction and dissatisfaction levels;
- Follow a strict regimen of hardware and software design reviews;
- Track first pass test yields and returned product rates;
- Carry out an intense program of vendor/subcontractor evaluation;
- Perform an extensive activity in corrective and preventive action and customer complaint response.
- Closely monitor our financial goals.

A more detailed discussion of quality objectives is found in Section 5.4.1 of this manual.

Communication of the Quality Policy and Its Status Growth uses several means to propagate its quality policies to all employees. Aside from the monthly top management meetings, such methods include

- The assignment of champions who establish quality teams to document, implement, and demonstrate the continuing effectiveness of the quality management system (refer to Growth's champion summary maintained by the ISO 9000 management representative) Champions.Doc;
- Weekly business review meetings by each top manager with his or her staff;
- The monthly newsletter;
- Quarterly business presentations by the GEMT to all employees;
- Posting of the quality policy in key areas of the facilities;
- Highlighting statutory and regulatory requirements to employees through the newsletter, policies, and procedures, and as part of the quarterly GEMT presentation;
- Reviewing the quality policy in the same manner (refer to Section 5.3 of this manual);
- Presenting quality objectivesas part of the business plan (refer to Section 5.4.1 of this manual);
- Holding quarterly management reviews by the GEMT to examine progress against targets (refer to Section 5.6 of this manual);
- Providing resources by means of the business plan and constantly reviewing the operation in the series of meetings described (refer to Section 6.1 of this manual).

5.2: Customer Focus

The vice president of sales and marketing is responsible for customer satisfaction management. The method used to determine and enhance customer satisfaction and determine and minimize customer dissatisfaction includes customer surveys, customer service feedback, and feedback from the sales offices in the form of weekly reports (refer to Sections 7.2.1 and 8.2.1 of this manual).

5.3: Quality Policy

Growth Corporation's Quality Policy

The Growth Corporation (Growth) is committed, at all levels of the company, to total customer satisfaction. To meet this commitment, we provide products and services that fulfill customer expectations and provide quality at levels greater than what is available from any of our competitors.

Our quality management system is based on the ANSI/ISO/ASQC Q9001-2000 international standard and is supplemented by the applicable clauses of the ANSI/ISO/ASQC Q9000-3-1997 computer software guidelines.

Growth is fully committed to continually improving the effectiveness of our system by means of constant top management review and oversight. This level of attention is complemented by formal management reviews in which the QMS is reviewed for continuing suitability, internal audits, extensive training, and an aggressive corrective and preventive action program that includes cross-functional teams for root-cause analysis and problem resolution.

To ensure the integrity of our system, quantitative quality objectives based on operational metrics are established, monitored, measured, and reviewed by managers who are held accountable for their results. In addition, all of our employees are thoroughly trained in our quality policy and quality management methods and are supplied with the resources required to ensure that such methods are effective.

At Excellent, business objectives and quality objectives are synonymous.

Signed: *Fran Dewolf* Dated: January 1, 2002
President and CEO
The Growth Corporation

Quality Policy Communication The methods used by Growth to communicate the quality policy to all employees is covered in Section 5.1 of this manual.

5.4: Planning

5.4.1: Quality Objectives

As previously mentioned, the GEMT formulates quality objectives, their metrics, and targets as part of the annual Growth Business Plan. We also listed a number of business objectives that were required to satisfy our mission statement. In this section, we will quantify these primary objectves and add additional objectives that supplement that set. The establishment of primary objectives and support objectives is the means by which quality objectives flow down through all pertinent Growth functions. Table 22.1 is a sample of some of the critical objectives set by Growth. The complete set of business/quality objectives can be viewed through this icon. 🗁GBP.Doc

5.4.2: Quality Management System Planning

Quality Planning As previously discussed, Growth's president and CEO is responsible for the annual development and publication of the business plan. The business plan is the framework within which the quality objectives, quality plans, quality policies, and changes to such activities are formulated as part of Growth's continual improvement directives. To ensure the overall integrity of the GCQMS, all QMS changes are subject to review and approval of the GEMT and become an integral part of the business plan. Section 7.1 of this manual provides more detail on this subject.

Corporate and Interdivisional Interfaces The Growth Corporation interfaces with other Stable Corporate facilities. In those cases where Growth provides services to other facilities, the transactions are performed exactly as if Growth were selling its services to a customer. In those cases where Growth receives services from another facility, the transactions are performed as if Growth obtained material from a vendor or subcontractor. In addition, the functions of financial analysis and information technology are shared directly with Stable. In such cases, the transactions are covered in procedures controlled by Growth.

5.5: Responsibility, Authority, and Communication

5.5.1: Responsibility and Authority

The following organizational chart (Figure 22.4) indicates the functional relationships of all personnel in Growth along with the indicators (Figure 22.5) that summarize which employees have the responsibility and authority to

Table 22.1
Growth's Quality Objectives Matrix

Marketing and Sales Primary and Supporting Quality Objective(s)			
Primary: maximize customer satisfaction and minimize customer dissatisfaction levels			
Metric	Target	Champion	Intranet location
Percentage of customers who reorder per year	100%	Vice president of sales and marketing	SandMreorder.xls
First support objective: survey customer opinions on overall performance			
Metric	Target	Champion	Intranet location
Number of returned surveys versus surveys mailed	>50%	Manager of direct sales	SandMsurveys.xls
Design Engineering Primary and Supporting Quality Objective(s)			
Primary: on-time hardware and software design reviews			
Metric	Target	Champion	Intranet location
Percentage of design reviews versus plan per project	100%	Vice president of design engineering	DesignReview.xls
First support objective: projects completed on time against plan			
Metric	Target	Champion	Intranet location
Percentage of project completed versus plan per project	>80%	Project engineers	EngProjects.xls
Manufacturing Primary and Supporting Quality Objective(s)			
Primary: ship product as specified by the customer-agreed-to shipping date			
Metric	Target	Champion	Intranet location
Percentage of shipments that meet ship date	>95%	Vice president of manufacturing	Shipping.xls
First Support Objective: Reduce NCMRs in assembly			
Metric	Target	Champion	Intranet location
Number of NCMRs per product line	Zero	Assembly supervisors	NCMRs.xls
Second support objective: optimize first pass yields			
Metric	Target	Champion	Intranet location
First pass yields per product line	80%	Vice president of manufacturing	Product Yields.xls
Third support objective: optimize vendor/subcontractor evaluation on-time deliveries			
Metric	Target	Champion	Intranet location
Vendor percentage on-time deliveries	>98%	Purchasing supervisor	VendorsOT.xls
Fourth support objective: optimize response to nonconformities			
Metric	Target	Champion	Intranet location
Response time to resolve nonconformities	Minimal	Vice president of quality assurance	NCRTime.xls

Table 22.1 (continued)

Customer Service Primary and Supporting Quality Objective(s)			
Primary: minimize returned product rates			
Metric	Target	Champion	Intranet location
Percentage of returned goods versus total field population per product line	<1.0%	Vice president of manufacturing	ReturnRate.xls
First support objective: optimize final test protocols			
Metric	Target	Champion	Intranet location
Percentage of critical functions tested	100%	Project engineers	CriticalTests.xls

Figure 22.4 Organizational structure of growth.

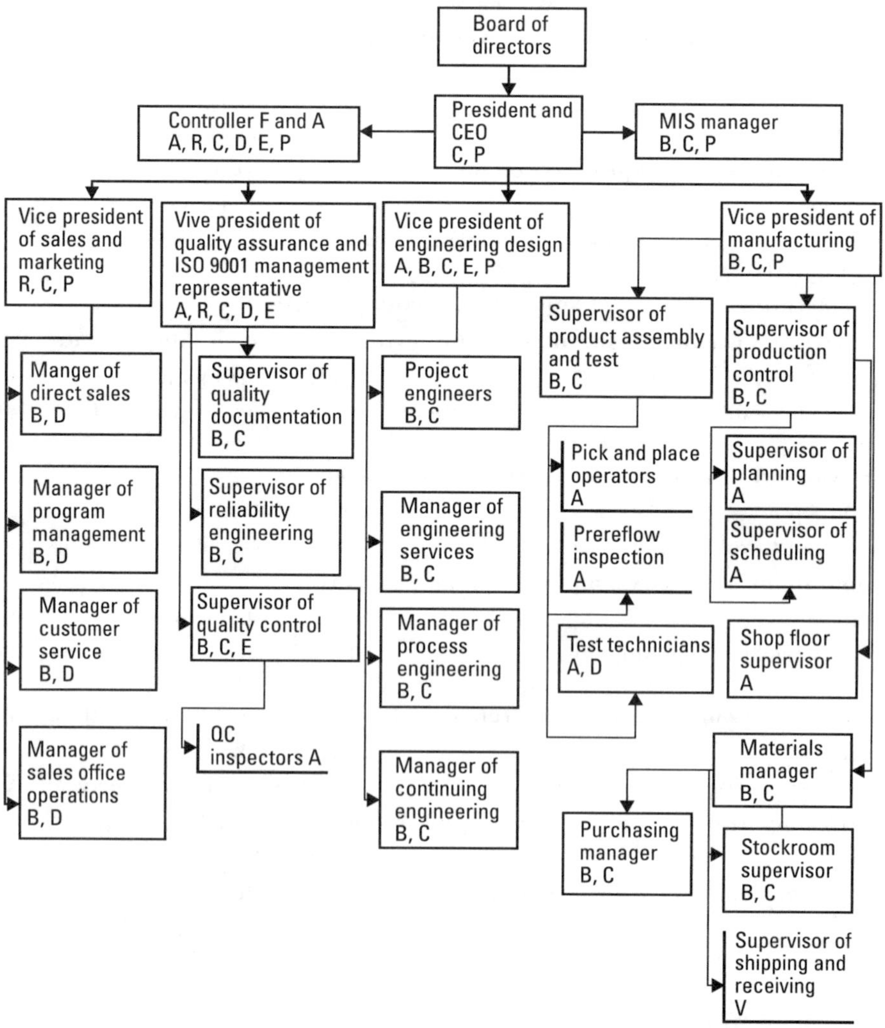

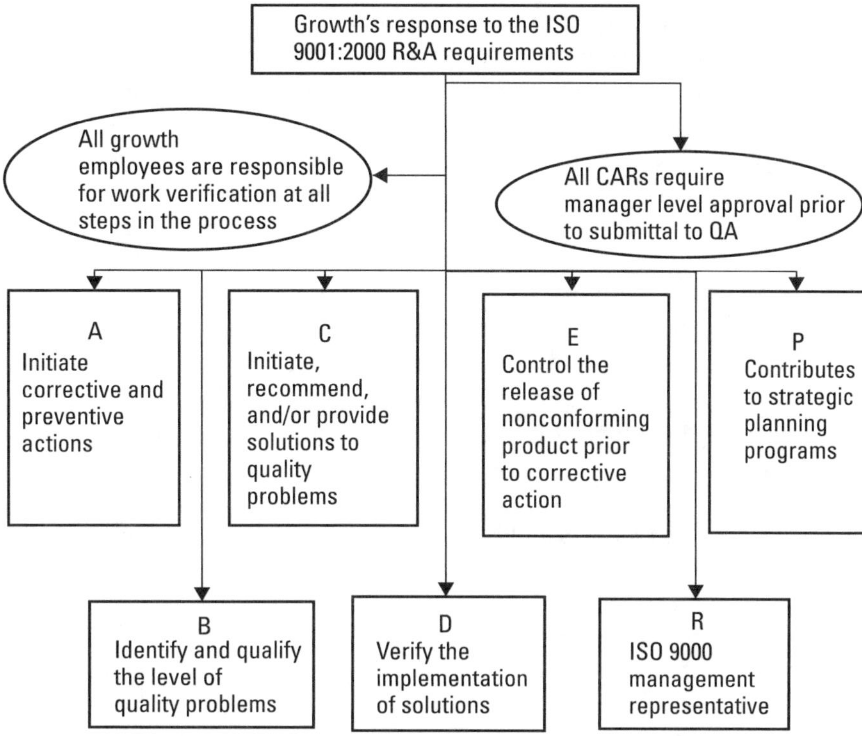

Figure 22.5 Responsibility and authority legend for growth.

manage, perform, and verify work-affecting quality. The organizational structure and indicators are posted in the cafeteria and presented during the quarterly GEMT presentations.

With regard to verification, all employees of Growth are required to constantly monitor their work to ensure that all quality requirements have been satisfied.

In addition, as indicated in Figure 22.1, the process champions also serve as ISO 9000 stewards, where each steward is responsible for the effective documentation, implementation, and demonstration of effectiveness for their processes.

President's Direct Report Management Responsibilities

President and CEO Responsible for developing and overseeing the execution of Growth's annual operating budget and managing the successful implementation of that budget. As the top executive, the president is required to articulate Growth's long-term strategies and to serve as the primary interface between the corporation and the board of directors.

Vice President Sales and Marketing Required to achieve the annual domestic and international sales quotas and direct the efforts of the sales personnel, program managers, customer service, and the Washington, DC, and Seattle, WA, sales offices. The position also requires the creation and implementation of the marketing plan, which includes media contact, direct mail, trade show activities, and OEM contacts.

Vice President of Design Engineering Serves as the chief engineer for Growth and is responsible for the development of the company's product lines. This activity includes synthesizing customer requests for product development and the general oversight of all aspects of hardware and software design functions. Process and continuing engineering as well as engineering services, bid control, and engineering documentation are directly controlled through this position.

Controller Creates and generates all financial information for internal and external users and is responsible for financial planning and management of the accounting functions. The controller is the primary interface with the chief financial officer at Stable.

Vice President of Quality Assurance Serves as the ISO 9000 management representative for Growth and manages all phases of quality assurance, quality control, and reliability engineering. The position also requires the oversight of the policy, process, procedural, and format document control for Growth.

Vice President of Manufacturing Responsible for the management of the production, materials control, and all purchasing for Growth. This position includes the functions of inventory management, the evaluation of vendors, and the support of prototype product builds.

MIS Manager Responsible for the design and implementation of all phases of the information technology system at Growth and is the primary contact with the corporate information technology officer (CITO) at Stable. The MIS manager controls the day-to-day effectiveness and backup systems for the division.

5.5.2: Management Representative

Appointment The president of Growth has appointed the vice president of quality assurance to continue as the ISO 9000 management representative during and after the 2000 upgrade effort. Notice of this appointment was effective on the first day of February, 2001, and distributed to all employees of Growth via the cafeteria posting. Such duties remain in addition to the usual activities of the vice president of quality assurance.

Duties of the ISO 9000 Management Representative In this position, the representative has the authority to establish, implement, and maintain an effective ANSI/ISO/ASQC Q9001-2000 quality management system within Growth, subject only to the review of the GEMT.

The duties of the ISO 9000 management representative include the establishment and management of our plan to meet the established upgrade certification timelines and to provide any necessary direction to the ISO documentation teams. The representative reports to the GEMT at the management review meetings on the status of the QMS in the areas of at least

- Progress against our goals and objectives;
- Internal quality audits;
- Corrective and preventive actions and customer complaints;
- Training;
- The state of documentation and implementation of the quality management system;
- Any need for QMS improvement.

Thus, the appropriate actions can be taken to continually improve the system.

Registrar Interface The ISO 9000 representative also maintains close contact with our accredited registrar, and finalizes the dates for the off-site document review and on-site certification upgrade assessment.

Ensure QMS Integrity In addition, the representative ensures the efficacy of the QMS by directly taking part in the internal quality audit and corrective and preventive action programs within Growth. The representative uses the internal audit program as one means to check on the integrity of QMS processes.

Ensure Employee Customer Awareness The representative participates in new employee orientation programs to make sure that all employees are aware of Growth's customer requirements. To further enhance this effort, the representative publishes updates in the monthly newsletter on customer acceptance of Growth's performance and posts performance-against-targets graphs on the cafeteria post board.

5.5.3: Internal Communications

As we have previously discussed, to measure QMS effectiveness and continual improvement, internal communication at Growth is spearheaded through the use of quality objectives, metrics, and targets at all pertinent levels of the corporation. The flow of information throughout Growth is enhanced by highly effective engineering and manufacturing software applications programs on the Growth intranet, as well as the use of the cafeteria post board and the monthly newsletter. Other communication tools at employees disposal include e-mail and a series of weekly, monthly, and quarterly management-led review meetings.

5.6: Management Review

Quarterly Review

The ISO 9000 management review is held quarterly and is chosen from one of the monthly top management meetings attended by the GEMT. At this meeting, the total business performance is reviewed and the suitability, adequacy, and effectiveness of the ANSI/ISO/ASQC Q9001-2000 QMS is determined and the necessary actions taken to improve its performance.

Review Inputs The agenda for the review is set by the president and includes, but is not limited to, the following [the presentation of such data is by the appropriate attendees]:

- Total corrective and preventive action and customer complaints program, including an analysis of process performance and product conformity and nonconformities and customer feedback;
- Results of the total auditing program (first, second, and third party);
- Review of quality objectives as compared to plan;
- Currency of the training program;
- Establishment of opportunities for improvement (OFIs) that are then presented to action teams for resolution;
- Consideration of any need for changes to the QMS;
- Recommendations for improvement, especially as they relate to the quality policy and quality objectives;
- Report on follow-up actions from previous reviews.

Review Outputs The minutes for this meeting are written and maintained by the administrative assistant to the president. GEMTMinutes.Doc

The minutes contain, in addition to the report summaries, GEMT decisions and actions that consider:

- Methods to improve the effectiveness of the QMS and its processes for possible action team assignments;
- Possible product improvements based on customer specifications that could be assigned to design engineering;
- Possible redistribution of resources to enhance process and product performance for assignment to department heads.

Supplementary Reviews Monthly department manager reviews and weekly operational-level reviews are also held as a way to analyze the effectiveness of the system on a much finer grid than over 3 months. Minutes of such reviews are also maintained by the pertinent manager or supervisor and the key information collected at such meetings is funneled into the quarterly review, as appropriate.

Joint Software Reviews Growth's software development protocol requires a close interaction with its customers in the form of jointly reviewed conformance to customer specifications. Conformance is based on software acceptance testing at both Growth and the customer's facility. The vice president of design engineering schedules and manages this activity.

22.6 Resource Management

6.1: Provision of Resources

As part of the Business Plan, Growth composes and implements an annual operating budget at both the top management and second-tier management levels. This budget is maintained by the controller and is used to generate a personnel hiring plan, the capital spending plan, and detailed operating budgets for each department at Growth.

The purpose of the budget is to provide the necessary resources to continually improve and effectively implement and maintain the GCQMS through the efficient distribution of capital and personnel. When there are sufficient resources available to fully staff the organization, provide a vigorous training program, and procure manufacturing materials on a timely basis, the company dramatically raises the probability of a satisfied customer by shipping product that meets customer specifications. Growth's history of extensive repeat product orders is indicative of the efficacy of this process.

During the fiscal year, top management reviews the budget on a monthly basis and compares the division's performance to the plan. The plan is then adjusted as required. Key elements of this plan include an extensive training program for all employees, including all aspects of management, hardware and software development, and internal quality auditing.

6.2: Human Resources

General The GEMT annually reviews departmental needs and financial resources necessary for Growth's continued expansion. The president and staff jointly determine the financial budget of the training program. The training program is documented in the training program and job descriptions documents. 🕮Training.Doc and 🗀JobsDescript.Doc

Planning ISO 9000 training is an ongoing process accomplished through internal quality audits, specific training sessions, company and departmental meetings, and any specific external training as required. All other training plans are initiated as part of the business plan, which includes new employee orientation and auditor training. The purpose of training is to make sure that each employee is qualified for their position on the basis of their education, training, skills, and experience as specified, as appropriate in the job descriptions.

Qualification and Needs Analysis The departmental managers assess their departmental needs and create job descriptions when needed, which include education level, skills level, training, and work experience needed to perform specific job functions. In this manner, Growth ensures that all employees are fully qualified for their positions.

Job Description The job descriptions are created at the time of hiring and maintained and archived in the controller's office. Each job description contains the conditions for education, skills, training, and experience for the employee. The job descriptions are amended as needed upon meeting and interviewing job candidates.

Training All employees are trained as deemed appropriate by the department manager. Training is in accordance with the individual employee's education, skills, training, and previous work experience as well as Growth's quality commitment. The manager is available to tutor any employee, specifically for any job-related questions. The manager chooses from on-the-job training, or special internal or external programs.

Specific training programs include classroom work, where all engineering and operational employees receive in-house video training covering all aspects of manufacturing. This includes electrostatic discharge (ESD) awareness, OSHA health and safety requirements, proper hand soldering methods and techniques, through-hole (T/H) and surface mount technology (SMT) process assembly and inspection techniques, component identification, and technical terms and definitions.

The use of quarterly GEMT presentations with all employees and the posting of progress towards quality objective targets are several of the ways that Growth ensures employee awareness of their efforts and how their work contributes to the overall success of Growth. The monthly newsletter also helps in this regard.

Management and professional training (both hardware and software) are provided by on-the-job training or through company-funded courses deemed appropriate by the managers. The training period ceases when the manager deems the employee appropriately trained. Management strives to maintain a quality working environment for all employees at Growth.

Reviews To determine the effectiveness of training, each employee is reviewed annually by his or her manager to ensure the highest work quality. Discussions cover work quality, strengths, weaknesses, and areas of improvement for the individual and for the department. Previous reviews are maintained and archived. Reviews can be performed on a computer template or with a written or verbal structure. In addition, managers and supervisors are required to continuously monitor employee performance and recommend additional training as required.

Each new employee is reviewed by his or her manager at the a time period decided upon at the date of hire. The results of these reviews are archived in the individual employee's file located in administration. The manager and employee discuss any issues pertaining to the quality of the employee's work. If a manager deems it necessary that an employee requires additional training to improve the quality of his or her work, it will be discussed with the employee and provided for by Growth. All the training programs are provided and run by the department manager.

Records Hard-copy orientation and review records are maintained by the controller and ongoing training records are maintained in the individual employee files. These records consist of all present and previous employees and are archived in each employee's personnel files in administration. Records are maintained for a period of time at the discretion of the individual manager. A summary of each employee's training is maintained online. EmployeeTrain.Doc

6.3: Infrastructure

The vice president of manufacturing provides plant engineering functions for Growth. Such services include the efficient operation of the facility and presentations to the GEMT for additional workspace and repositioning of equipment and offices. The movement of any new process, software, or equipment onto the manufacturing floor requires the vice president's approval and is controlled by the ECO process. All activities of this type that require purchased capital equipment require a payback analysis. Software revisions also require the approval of the vice president of engineering design.

Preventive maintenance of machinery and fixtures are also included in this program. A full program of preventive maintenance ensures long-term reliable operation of the capital equipment. Maintenance is performed at varying frequencies, as required.

The materials manager manages any transport requirements and the MIS manager manages any communication requirements.

Infrastructure considerations are discussed in detail in Infrastructure Guidelines. 🗁InfrastructureGuide.Doc

6.4: Work Environment

Responsibility The manufacturing vice president is responsible for a clean and efficient workplace that is safe and comfortable for Growth's employees.

Work Flow The layout of the process area is designed to allow for an efficient and flexible process flow. Duplication of effort and retracing is minimized.

ESD Control ESD control is required throughout the manufacturing process and the implementation and training in ESD is the responsibility of the reliability engineering supervisor.

Environmental Controls The work environment is maintained to ensure that the processes are stable and employees are adequately comfortable. For this purpose, four automatic temperature control thermostats are located throughout the manufacturing area.

22.7 Product Realization

7.1: Planning for Product Realization

As previously discussed, Growth's president and CEO is responsible for the annual development and publication of the business plan, which is the framework within which the quality objectives, quality plans, quality policies, and changes to such activities are formulated as part of Growth's continual improvement directives.

Growth uses a hierarchical documented system of policies, processes, procedures, and forms to control the product realization activities. For example, software planning includes but is not limited to design documents, functional specifications, and quality documents to verify that the functional specification has been met. The process documents for this purpose are displayed in Figure 22.1.

Planning Output The product realization process is driven by both forecasts and directly received customer purchase orders. Output from this set of inputs is a combination of MRP reports and spreadsheet plans and schedules constructed by the manufacturing supervisors of planning and scheduling and approved by the vice president of manufacturing.

Procurement The acquisition of major capital equipment, new processes, and specially skilled employees is the responsibility of department managers and requires approval by the president.

Compatibility Growth employs design and documentation reviews to ensure that new products are manufacturable. A combination of program management/project management teams during design and the use of a continuing engineer during the production start-up phases ensure the effective transfer of new products into manufacturing. Process engineers are then used to maintain the production lines.

Installation Installation guides are provided to customers so that they can effectively install Growth products into their systems. Developer's guides provide information for developers writing software applications and software libraries such as microsoftware plus have resident software library references.

Updating Design engineering in conjunction with quality assurance is responsible for updating, as necessary, quality-control procedures, inspection and testing techniques, and the development of new test instrumentation.

Test Equipment Design engineering has the primary responsibility of designing and implementing any measurement equipment that exceeds state-of-the-art specifications. Such equipment must be available prior to production release and be clearly identified in the project plans.

Verification and Validation Design engineering is primarily responsible for the functions of verification and validation at scheduled stages in the product design plan. Upon release to production, quality assurance imposes verification and reliability activities on the product at all stages of manufacturing.

Standards and Codes and Workmanship Standards The Standards and Codes Procedure lists all of the regulatory and statutory requirements imposed on Growth's products [e.g., ISO 9000 Standards and applicable guidelines; institute for interconnecting and packaging circuits (IPC) workmanship standards; and the CE mark]. The procedure includes a master list that defines the standards and codes used; the responsible employee; how they are kept current; and where they are located. The internal design standards are also included in this procedure. 🗁Standards.Doc

Workmanship standards in the form of a series of IPC documents and reference manuals are provided to manufacturing. All manufacturing personnel are required to attend and satisfactorily complete a certification program based on IPC Standards.

Inspection and Testing Process Inspection and testing processes at Growth are created, controlled, and recorded by quality assurance and include reliability testing at accelerated temperatures over extended test times. The vice president of quality assurance is responsible for all inspection and test procedures at Growth. The primary document for this activity is contained in the document entitled "Inspection and Testing Processes." 🗁InspectTest.Doc

Product Quality Objectives and Requirements Refer to Table 22.1 for typical product objectives used to ensure a high level of product quality and performance.

Records Refer to Section 4.2.4 for a more detailed discussion of the records maintained by Growth to confirm that product performance meets customer requirements.

7.2: Customer-Related Processes

7.2.1: Determination of Requirements Related to the Product

Process The sales and marketing procedures are contained in the document entitled "Sales and Marketing Process." The vice president of sales and marketing is solely responsible for the content and accuracy of this document (inclusive of Section 7.2.2). 🕮SandMProcess.Doc

Determination of Customer Specifications In addition to sales to customers based on published brochures and price lists, sales and marketing program managers and design engineering project engineers establish direct contacts with customers in order to determine customer specifications. The functional specifications include all statutory and regulatory requirements. In those cases where the customer has inadvertently missed a specification that impacts form, fit, function, safety, or reliability, Growth negotiates such specifications into the functional specification.

For direct sales, a combination of sales orders and the customer's purchase orders are used to clearly define delivery requirements and warranty conditions. Customer service manages all after-sales activities, which includes returned goods and service contracts. Service contracts require the customer service manager's approval.

7.2.2: Review of Requirements Related to the Product

Forecasts The vice president of sales and marketing prepares a rolling monthly forecast for review with the top managers to ensure both hardware and software product availability and to meet contract or accepted order specifications.

Standard Products The president has final review authority, and the vice president of sales and marketing publishes Growth's standard price list. PriceLists.Doc

Standard off-the-shelf products are quoted by the sales staff by discounting the published price lists. Any nonstandard discount requires the approval of the vice president of sales and marketing.

Custom Products Custom quotes and customer contracts that require special product or pricing changes are reviewed and approved by the GEMT to ensure that Growth has the proper financial, marketing, engineering, quality assurance, manufacturing expertise, and capacity to take on the project.

Review of Customer Specifications Growth's sales staff, either at corporate headquarters or in the sales field offices, is required to review all written and verbal quotes with the customer and/or prospect base to ensure that customer specifications are clearly addressed.

New Products New opportunities for products—either hardware or software, solicited either by internal referendum or from the external market—are reviewed by the vice president of sales and marketing and presented to the GEMT to make the final decision on acceptability. The vice president of design engineering cannot accept a new project without this review and approval.

Verbal Orders Before a verbal purchase order can be accepted, the order must be documented by the sales staff, and, as with any written purchase order, the staff must ensure that the product specifications and pricing are correct.

Conflicts Growth's sales staff, both corporate and in field sales, have the responsibility and authority to resolve any issues in contracts, to resolve order discrepancies, and to raise such issues to whatever level of authority is required.

Amendments Sales and marketing personnel have the sole authority to amend orders regarding changes to written contracts. They must notify all of the affected departments in writing and manage any subsequent activity, particularly when a new product effort is involved. In all cases of amendments, a new sales order is opened.

Records Quotes are dated and stored by the sales staff. If a sales manager for a specific region is unavailable, another member of the staff has the authority to resolve any issue that might arise. All pertinent verbal and written correspondence with Growth's customers is dated and stored locally. Quotes are filed electronically. Records also include (but are not limited to) sales orders, customer purchase orders, sales acknowledgments, memos, and customer specifications.

7.2.3: Customer Communication

Product information is supplied to customers by means of product brochures and advertising managed by the vice president of sales and marketing. The manner in which inquiries, contracts, order handling, and contract amendments are managed is covered in Section 7.2.2 of this manual.

Customer feedback is obtained via customer surveys coordinated by the manager of direct sales; returned goods analysis, by the manager customer service; and management of customer complaints, by the vice president of quality assurance. The specifics of customer-complaint analysis are to be found in the quality-assurance manual. 🕮QAManual.Doc

7.3: Design and Development

Design Flow Growth's overall product design protocols cover six stages, and this concept is shown graphically in Figure 22.6. 🕮DesignEngProcess.Doc

Guidelines Growth maintains documented and control procedures throughout the life of the product. A detailed description of each stage in the process is located in the hub document entitled "Hardware and Software Development Guidelines." 🗀H/WandS/WGuidelines.Doc

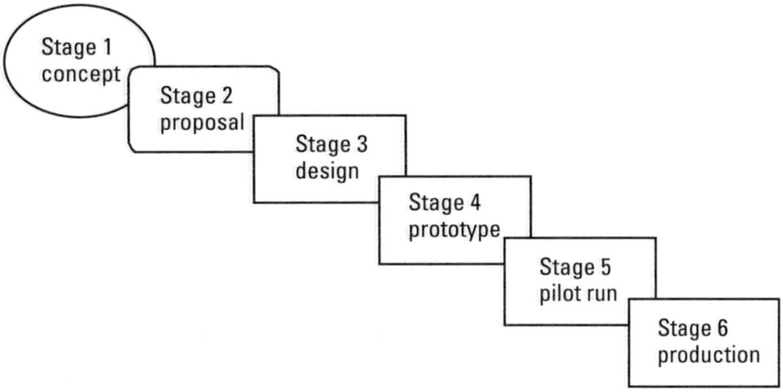

Figure 22.6 Product design guidelines.

7.3.1: Design and Development Planning

Product Management Products are designed under a program manager and project engineer protocol. The program managers are assigned by the vice president of sales and marketing and the project engineers are assigned by the vice president of design engineering. The project engineers are assigned to programs based on their expertise in hardware or software and each project engineer is required to form a cross-functional program team and specify clearly the authority and responsibility of each team member.

Program Plans The team creates the program plan, and the program manager is responsible for all administrative activities, which includes maintenance of the program plan and the program files. The vice president of design engineering in the role of chief engineer approves the final plan after an initial design review. Each plan includes specific activities related to design review, validation, and verification of both hardware and software tasks.

Communication The program manager schedules weekly team reviews to monitor the use of program resources and to maintain a companywide perspective on the product's development. It is common to invite technical specialists from outside the team to help resolve design issues. Variances from plan are elevated to the chief engineer who either approves the variance or requires that an appropriate corrective action be taken.

Resources The vice president of design engineering responds to requests by the program manager for the necessary engineers, equipment, facilities, and support personnel.

Updating The program manager updates each design plan periodically based on design review, management reviews, or status meetings.

7.3.2: Design and Development Inputs

The program manager and project engineer are responsible for the functional specifications based on various internal department and customer inputs and on Growth's decision of the product definition. The functional specification includes all statutory and regulatory requirements (e.g., use of the CE mark) and requires the final approval of the chief engineer.

The product proposal results from this specification, and its scope and complexity is proportional to the program's cost. Any ambiguities or conflicts that result from the functional specification are brought before the chief engineer for resolution and for approval of completeness.

7.3.3: Design and Development Outputs

The chief engineer is required to ensure that the final functional specification agrees in detail with the customer's requirements and only then approves and releases the document.

Upon release of the final functional specification, the design team develops the user's manual, which contains instructions for any special handling procedures that may be required (e.g., ESD handling instructions) and product maintenance. In addition, diagnostic test procedures are developed by the project engineer to verify and validate the product prior to production handoff.

Additionally, the design team establishes the technical file, which is part of the transfer to production. The technical file supplements the design history file and contains (among other items) the bill of materials, production work instructions, servicing procedures, product acceptance criteria, and product characteristics essential for safe and proper use. This information is designed to enable purchasing, production, and customer service to appropriately manufacture and service the product.

7.3.4: Design and Development Review

The program manager schedules design reviews into the program plan based on the complexity of the program and keeps the minutes from these reviews in the design history files, which form a part of the technical file. Several types of reviews are scheduled, which include technical reviews to define design, software code, algorithms, and production and manufacturability, as well as marketing and sales reviews to ensure that the program will meet the customer's changing needs in a dynamic marketplace. Appropriate guests are added to the design team as required to cover specific technical and/or marketing issues.

The minutes specify both hardware and software problems, and team members are assigned to analyze and offer solutions to such problems. Each design review includes a progress evaluation of such assignments.

7.3.5: Design and Development Verification

Project engineers test all new products against the final functional specifications as a normal part of the project plan. In addition, all hardware products are verified before release into production by means of a final prototype build and test, to ensure that all supporting documentation for production is available and correct.

Software engineers continuously test, debug, and verify software code as a normal part of the design process. In addition, all software products are verified before release into production by means of a verification copy of the software to ensure completeness of the production transfer package.

The program manager and project engineer are jointly responsible for performing and documenting the verification process and its results and the completeness of the design history files in this regard.

7.3.6: Design and Development Validation

All hardware and software products are validated before release into production to ensure conformance to the final functional specification, which includes all customer requirements. This process requires a system-level test strategy using a typical customer's system. In certain situations, the customer may ask to be present and validate acceptance through their signature.

The responsibility for performing and documenting the validation testing is jointly held by the program manager and the project engineer, who are also required to include a complete record of the activities in the design history files.

Transfer to Production A continuing engineer moves with the project from pilot line runs into forward production, and then remains with the program until manufacturing engineering phases in. An ECO is used for this transfer.

7.3.7: Control of Design and Development Changes

Design changes that result in variations from the functional specification are reviewed by the chief engineer for approval. The change, or rejection of the change, is documented in the design history files by the program manager. The chief engineer has the discretionary authority to call a more general management review if required.

An ECO, maintained by the engineering documentation manager, is used to release products to production, and once the product is released any changes are made via the ECO process.

Part of the ECO function, which contains a history of actions taken to assess change validity, is to make certain that all changes have been properly reviewed, tested to verify the efficacy of the change, and validated against customer-specific requirements, as appropriate, particularly with regard to the impact on component parts and product already in the field.

7.4: Purchasing

7.4.1: Purchasing Process

The purchasing documentation includes a set of procedures and specifications required when procuring subcontractor services, piece parts, and noninventory and other items. This information is contained in a controlled three-ring binder entitled "Purchasing Manual" maintained by the purchasing manager. 🕮PurchManual.Doc

Incoming Inspection All raw material received from vendors is verified by receiving and sampled by quality control to ensure conformance to specification. Product that does mot meet specification is placed in material review board (MRB) for disposition. In addition, printed circuit boards require a certificate of conformance by the contract vendor with deliveries. In cases of immediate or urgent production requirements, a waiver procedure is used to identify material released for production prior to the issuance of a complete documentation package.

Evaluation of Subcontractors Growth initially selects its subcontractors on the basis of site visits, questionnaires, quotations, and references. In all cases, a long-term commitment to quality assurance, especially in terms of ISO 9000, is sought with its key customers.

After selection, site quality audits when appropriate are conducted jointly by the quality-assurance department and manufacturing department to maintain an ongoing partnership relationship. In addition, periodic evaluation reports, which inform the vendor of its progress against on-time delivery, performance, and quality, are published by the purchasing manager.

Printed circuit board design houses are specifically evaluated by the vice president of manufacturing and the vice president quality assurance. Certificates of compliance and/or analysis are required from the printed circuit board design house.

The purchase of noncritical components is entirely at the discretion of the buyers.

The records of approved subcontractors are maintained in the approved vendor list (AVL) maintained by the purchasing manager. Subcontractors are either added or removed from this list through a continual evaluation process documented by the purchasing manager.

7.4.2: Purchasing Information

Growth's purchase orders (POs) include the PO number, the date the order was placed, the PO type, the date the PO was last changed, the quantity ordered, the part number and its description, the vendor, the price, and the required quality standards, when appropriate. The quality requirements of both material and personnel are supported by attached engineering drawings and specifications.

The POs are controlled in a numbering sequence that is kept in logs controlled by the buyers. Two different types of POs are used to purchase either inventory or noninventory items. All inventory POs require the signature of the purchasing manager prior to release. Noninventory orders are signed off by the buyers.

7.4.3: Verification of Purchased Product

Growth performs source inspection, when appropriate, at the subcontractor's facility. In those situations, the purchasing manager is required to formally alert the vendor prior to the visit and develop a protocol that establishes the conditions under which the source inspection can be terminated and future shipments can be released.

If agreed to contractually, Growth allows its customers—or their representatives—to inspect its product at either Growth or at the subcontractor's facility.

Growth neither uses the positive results of such an inspection as evidence of its effective vendor management, nor feels that it is a means of releasing Growth from always supplying acceptable product to its customers—nor that it affects the potential rejection of such product by the customer.

7.5: Production and Service Provision

Growth's manufacturing process is based on 10 stages as graphically demonstrated in Figure 22.7 (refer to the process document entitled "Hardware and Software Manufacturing Processes"). MnfgProcess.Doc: Customer Service documentation is found under this icon: CustSvc.Doc:

7.5.1: Control of Production and Service Provision

Production Control The vice president of manufacturing oversees the functions of production and materials control. The production control manager and the materials manager analyze the sales forecast and inventory status and produce production schedules that are approved by the director. The schedules drive the procurement process. Floor documentation released by the control team includes a process control document, a color-coded assembly drawing,

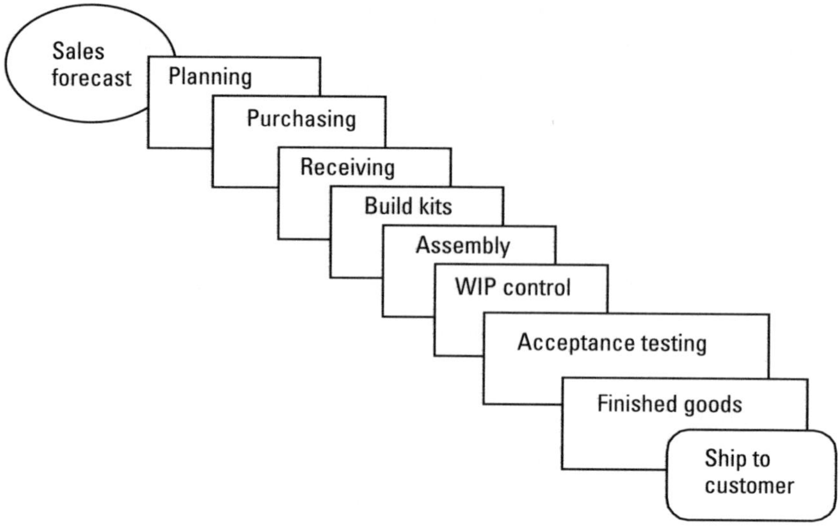

Figure 22.7 Product manufacturing process.

required inspection visual aids, and special mechanical drawings as required. Any special characteristics of the product are included in this documentation set and coded as necessary.

Continuing engineering makes sure that production equipment meets the product manufacturing requirements with respect to both accuracy and precision. The vice president of manufacturing is responsible for capacity and throughput requirements and authorizes the movement of processes and product-related manufacturing equipment onto the production floor.

Monitoring and Control Growth uses a series of software-based tests that display various pixel image patterns for visual definition, and test procedures are developed by design engineering for specific product lines.

The controlling process document is the process control document (PCD) (PCD.Doc), issued by engineering services for each assembly built. This document contains both customer documentation and Growth's process revision control system to ensure that the current product is built to the correct revision. This document fully defines each process step that an assembly must undergo. These documents are prominently displayed in the materials, surface mount, through hole, quality control, and other areas as required when an assembly is in process. This document also identifies all required tooling, programming, and special instructions as necessary. In addition, any component substitutions or special material preparation is included.

The PCD document requires sign off by materials, process engineering, manufacturing, quality, and when appropriate test supervision. The revision levels are controlled and updated under the ECO process.

Each process step is monitored and, when applicable, statistical techniques are used to measure process variables to ensure that the process is in control. Customer quality requirements have absolute precedence.

Workmanship Functional test suites are run against all hardware products to ensure that the products are functionally correct. Video quality is verified by running a series of software tests. IPC standards are used where applicable in the assembly process.

Nonrepetitive Processes Nonrepetitive processes, such as prototype builds, depend on very close customer interfacing to define levels of process controls. Special instruction sheets are issued as part of the process control document, when applicable.

Records and Nonconformances Any material or assembly found in nonconformance is promptly identified and segregated as such. Prompt action is then taken to bring the material back into conformance or disposition according to the protocols for the control of nonconforming product as discussed in Section 8.3 of this manual. The records of acceptance and/or rejection, which clearly show the responsible inspection authority, are maintained in their respective areas (i.e., either in materials or quality-assurance files).

Notification of products on hold is identified by a QC verbal communication to all affected areas. Any changes required to remove the product from hold is done through the ECO process.

All product that is shipped from Growth meets all required specifications. The shipper is required to verify the inclusion of all product and related components into the shipped package, including software and documentation as noted on the pick ticket for the order. The shipper validates the pick ticket with date shipped and their initials.

Software For required software, the software files to be replicated are revision controlled by means of the ECO process and its associated part number. Error checking is built into the replication process.

Records All records are maintained by either quality assurance or materials, as appropriate.

Customer Services Under the direction of the customer service manager, Growth provides various service functions to its customers, which include the following:

- Warranty repair;
- Out-of-warranty repair;

- Extended service contracts (hardware and/or software);
- On-site warranty support;
- Custom hardware or software services;
- Field implementation of ECO procedures for software and hardware;
- General servicing.

Warranty Repair Growth provides warranty terms and conditions as part of the general sale. Warranty repairs and mandatory ECOs are handled through the return material procedure (RMA) procedure.

Out-of-Warranty Repair Repairs or ECOs to hardware or software products that are out of warranty are handled through the RMA procedure.

Extended-Service Contracts At the discretion of sales and marketing, Growth offers extended-service contracts that may be for hardware or software products.

On-Site Support Reporting, verification, and tracking of on-site support issues are the responsibility of the customer service manager. Service reports are in the form of memoranda and are completed on site.

Spare Parts When appropriate, Growth supplies spare parts to minimize customer down time. The materials department is responsible for the activity.

Custom Services Custom services above and beyond Growth's normal servicing policy may be negotiated in an individual contract with the approval of the director or sales and marketing.

Field Implementation of ECO Procedures The customer service manager oversees the implementation of ECOs that affect hardware or software product in the field. At the discretion of the customer service manager, field exchanges in accord with the RMA procedure may be used to implement ECOs.

General Servicing The general servicing that Growth provides is defined as verbal and written technical support, historical and analytical information, and other assistance in resolving technical issues.

7.5.2: Validation of Processes Production and Service Provision

Special production processes, such as conformal coatings, are closely monitored and conform to strict qualifying procedures as defined by customer requirements. Records are maintained by the vice president of manufacturing and are related to the qualified process used, the qualifications of the personnel, and the qualification of specific equipment used. Revalidation of the processes and procedures as well as requalification of personnel is performed either annually when there is continuous production or as required for short runs.

In addition, because image quality is measurable on both a quantitative and subjective level, acceptance testing of product is performed by qualified test operators using calibrated test equipment. As with the special production processes, pertinent records are maintained by the vice president of manufacturing for this purpose.

7.5.3: Identification and Traceability

Growth uses product identifiers (PIs) to define the sales and marketing product name that may not be the same as the part number. The vice president of sales and marketing maintains a list of the approved PIs. All hardware and software documentation for a given product clearly states the PI.

The vice president of manufacturing is responsible for determining the serial numbers of all hardware and software products, and all products are labeled with either the PI or the part number and its revision level.

All hardware products are labeled with their serial number, and all software products are labeled with their release date. Software products are identified and traceable by part number and revision level. The vice president of manufacturing maintains a database that tracks board history.

Hardware documentation includes information needed to trace and record all printed wiring assemblies, box, and system-level products by board type and manufacturing lot. This information includes BOM, build, and test information.

A manufacturing resource planning (MRP) system is used to identify and track the progress of units as the lot moves through the manufacturing process. All product is followed and tracked by job number. Job process tags are attached to all product for use in identifying, tracking, and tracing the product through the factory. Each operation in the process is verified by the operator's initials. Different colors are used to differentiate standard product from returned product.

Inspection and Test Status When Growth receives components, printed circuit boards, and assembled materials, the receiver verifies and dates the packing slip. Upon receipt, quality assurance verifies the assembled material's inspection and test status.

Job tags that define job numbers and individual board serial numbers are used to monitor manufacturing and inspection process traceability on all boards processed.

All manufacturing and test processes are initialed and dated on the tag by the operator performing the specific process. All inspection processes are initialed and dated on the tag by the inspector performing the specific inspection.

Inspection status is further supplemented with the use of unique inspection stamp markings directly onto the surface of the printed circuit boards. The stamps use an indelible ink that was selected to withstand all cleaning processes. Either inspection or test stamps are marked on the assemblies as required before shipment.

The status of software is determined by the presence of the label that includes the appropriate part number and revision. Software products clearly indicate revision levels in accordance with documented numbering methods defined in engineering document control procedures.

Growth's in-house quality test records that define the hardware status are maintained in the first pass yield database. This database tracks boards by the product name, build lot, and serial numbers. The Growth RMAs database is used to record any nonconformance of both hardware and software products.

Physical locations in the production area for work in process and finished goods are also used for identifying the status of assembled product. Rejected or failed material is marked with a nonconformance color-coded tag, specific to the type of nonconformance. Products are required to be fully tested and burned in prior to shipment. If a product is shipped that does not meet full specification, a mutual agreement is established between sales and the customer. A hold tag is used to identify any product that may be placed on hold for any reason. Any subsequent changes to product are identified via the ECO process.

7.5.4: Customer Property

Customer-owned (supplied) materials at Growth consists of RMAs and miscellaneous engineering test systems and development equipment used in the design validation process. The RMAs are controlled by the RMA procedure and tracked by customer service that has the primary responsibility of responding effectively to customer returns.

Customer service also provides the required management for field exchanges, loaners, or evaluation units that are handled in the same manner as product returned for repair or refurbishment.

Miscellaneous test systems and development equipment is maintained and tracked by local area engineering managers.

Any customer-owned equipment or material that is lost or damaged in any way is reported to the customer service manager for disposition and corrective action with the customer.

7.5.5: Preservation of Product

Handling All Growth employees are trained to follow Growth's ESD practices and the effects ESD can have on Growth's products. Quality assurance is responsible for the maintenance of the ESD equipment and documentation. The adherence to these policies is the responsibility of the department managers and all employees.

Storage Receiving and quality assurance receive and approve all products that are shipped into the facility. Once the products are verified and received, they are put into a secured stockroom for storage. The stockroom supervisor is responsible for handling the transactions and activity in and out of the stockroom to prevent any damage or misplacement of components.

Parts and finished goods leaving the stockroom area must be signed out on a sign-out worksheet. The stockroom is organized by Growth's part numbers. The inventory control supervisor monitors these components on a regular basis.

Packaging Procedure Growth has a specific shipping procedure that must be followed when shipping customer products. The manufacturing process directs the shipper to the necessary documentation and equipment. The manufacturing processes describe the specific packing material that is used. These materials include static shielding bags for all board-level product. When requested, Growth packages product according to special customer specifications. The materials being used for packing are monitored by the shipper/receiver and the appropriate buyer.

Preservation There are separate areas designated for finished goods, works in progress, raw components, customer-owned material, and MRB. These areas are closely monitored by quality assurance. Growth adheres to shelf-life requirements where applicable.

Delivery Growth uses a segregated secured stock room for finished products that have passed final test and inspection. These products are available to ship for customer orders. Products are shipped free-on-board (FOB), the Growth Corporation, St. Louis, MO. When the customer does not specify a delivery method, Growth uses the most expedient, cost-effective, and quality-assured method. If there is a contractual agreement, Growth extends its protection responsibility to include delivery to destination

7.6: Control of Monitoring and Measuring Devices

Monitoring and Measuring Device Protocol Growth requires the calibration and maintenance of equipment used to either make absolute measurements or accept or reject product, e.g., oscilloscopes, multimeters, photometers and temperature probes according to Growth's Metrology Manual. Only inspection, measuring, and test equipment (IM&TE) used for monitoring and measuring processes and product, that requires calibration, are so marked. 🕮Metrology.Doc

Testing/Software Test protocols are initially designed by design engineering. They are released as part of the release package to manufacturing, where they are maintained under ECO control. All software is validated before use by means of both golden boards and diagnostics, and the software integrity is maintained under ECO control.

Testing/Hardware Test hardware used by Growth for development and production is maintained and/or calibrated by design engineering.

Measurement Uncertainty Calibrated equipment has documented tolerances. Where applicable, measurement uncertainty is determined by design engineering.

Records Calibration information is maintained by the director of quality assurance. It includes the definition of nationally and/or internationally recognized standards, equipment type, unique identification, frequency of calibration, calibration method, acceptance criteria, and actions required if equipment becomes uncalibrated. This also includes any calibration or maintenance falling outside the specified calibration intervals.

Technical Data If contractually required, Growth makes available to a customer all technical data related to the specific measurement equipment.

Control Responsibility Either the equipment manufacturer or design engineering is responsible for determining the accuracy and precision of any purchased testing or measurement system. The required accuracy and precision of the equipment is established either as a part of the selection process or when otherwise specified by the customer.

Calibration Calibrated equipment is visibly tagged or labeled, indicating the last calibration date and expiration date. This label also has the authorized signature of the person that performed the calibration.

Calibration information is maintained by the director of quality assurance. It includes the definition of nationally and/or internationally recognized standards, equipment type, unique identification, frequency of calibration, calibration method, acceptance criteria, and actions required if equipment becomes uncalibrated. This also includes any calibration or maintenance falling outside the specified calibration intervals.

Standards All calibrated equipment is calibrated against international or national standards, as appropriate. In some cases, internally created test protocols are used as test standards based on actual applications. Calibration plans are managed via logs that are maintained by quality assurance to indicate calibration cycles and frequency.

Calibration labels are used on all required IM&TE to alert operators that calibration is adequate or due. If calibration is overdue, operators are to immediately alert quality assurance and suspend use of the equipment until calibration is completed.

All equipment is sent out for independent calibration to companies selected by their capability with regard to using appropriately known standards. Quality assurance maintains logs of all of these transactions. A Paradox database file, CALIBRAT.DB is maintained, which lists calibration status for all equipment on a calibration cycle.

Invalidation Any product that has been validated with uncalibrated equipment is subject to a documented joint review by quality assurance and the director of manufacturing. The corrective action protocols are used when required. However, it is the responsibility of each operator to check equipment calibration status prior to each measurement. Appropriate actions are taken to correct any situation in which measurements were made with equipment found later to be out of calibration. Such actions include notifying the customer, retesting product, product recall, waivers, and rework.

Conditions All IM&TE requires room-temperature operation only and no special handling other than normal maintenance as prescribed in the equipment's operation manual.

Safeguarding Calibration labels and/or seals are placed in an appropriate location to prevent adjustments. If the label/seal is broken, the calibration becomes invalid, and the equipment may not be used until recalibrated. Operators are not allowed to make any adjustments to equipment. All adjustments are under the control of quality assurance.

Subcontractors IM&TE used by Growth's assembly houses, when required, is consigned to the house and maintained using Growth's calibration processes.

22.8 Measurement, Analysis, and Improvement

8.1: General

Monitoring and Measuring As part of the program to continually improve GCQMS effectiveness, an extensive program in monitoring and measuring, such as inspection and testing protocols at Growth, are created, controlled, and recorded by quality assurance and include reliability testing at accelerated temperatures over extended test times. The director of quality assurance is responsible for all inspection and test procedures at Growth. The primary document for this activity is contained in the procedure entitled "Inspection and Testing Processes." IandT.Doc

Analytical Methods Sampling techniques for incoming raw materials are performed by quality assurance against historically based sampling plans that include 100% sampling. Key materials such as printed circuit boards and other integrated circuits are either pretested before receipt or received with certificates of compliance. The rejection rate on some material (e.g., cables) is historically minimal, and a dock to stock process is used.

Growth uses data analysis and graphical techniques, such as Pareto charts, to study both the results of corrective and preventive action and the results of reliability testing and nonconforming product behavior. A limited program in statistical process control (SPC) is also in use, and, if successful, will be expanded at the discretion of the vice president of manufacturing.

In the case of software analysis, field feedback and failure data is collected and maintained by the vice president of quality assurance in conjunction with the software design manager for review and prioritization by the program managers and the customer service manager.

Procedure The document that describes Growth's statistical techniques is contained in the procedure entitled "Data and Statistical Analysis Processes." StatTech.Doc

Yield Analysis Production records the results on a board-by-board basis of first pass testing. The results are archived in an Excel database and used to plot a first pass yield. Analysis of yield information is the responsibility of the director of quality assurance. Yields.Doc

8.2: Monitoring and measurement

8.2.1: Customer Satisfaction

As discussed previously in Section 5.2 of this manual, the vice president of sales and marketing is responsible for customer satisfaction management. The method used to determine and enhance customer satisfaction and determine and minimize customer dissatisfaction includes customer surveys, customer service feedback, and feedback from the sales offices in the form of weekly reports.

This theme is further developed in Section 5.4.1, where we discuss the sales and marketing quality objectives in which customer satisfaction is highlighted. Additionally, repeat business is seriously considered to determine our customer's perception of Growth's quality performance. Some of the larger OEM customers send periodic reports that stipulate quality indexes against which we have successfully performed.

8.2.2: Internal Audit

Procedures Growth's procedure for internal quality audits is entitled quality audit processes. All audit protocols are managed by the director of quality assurance. 🗁Audits.Doc

The document details the following:

- Selection, training, and proficiency requirements of internal quality auditors;
- Way in which auditor independence is ensured by the controller;
- Gathering of specific information concerning the area to be audited by the auditors;
- Use of the auditor's checklist that is an Integrated ISO 9001 and ISO 9000-3 Standard and Software Guideline, respectively;
- Observance and testing of quality documentation and quality activities;
- Documentation of corrective actions for any noncompliances found by means of a corrective action report (CAR);
- Publication of the audit report;
- Follow-up by the lead auditor(s) to ensure the corrective actions have been completed and are effective in achieving the quality goals and objectives of Growth.
- The management of customer, vendor, and third-party audits.

Schedules The schedule of internal quality audits is determined by the vice president of quality assurance. For this purpose, an annual timeline is published, as all areas of the company are audited on an annual basis against the appropriate element(s) of the integrated standard.

The frequency of areas audited is based on their recent performance, CAR history, and the importance of that area within the quality system.

Records The results of the internal quality audits, recorded on internal quality reports by the auditors, are maintained by the vice president of quality assurance.

Corrective Action The corrective action items documented during the internal quality audits are reviewed by the auditor with the assigned company employee directly responsible for the quality activity. Mutually satisfactory date(s) are set for completion of the noted corrective actions.

Follow up The follow-up audits are conducted on the scheduled date(s) for completion of either action items or corrective actions. The results are presented by the auditor to the employees directly responsible for the quality activities audited.

Closure The action items or corrective actions are closed when the auditor determines that the actions taken are effective. This process may require several iterations.

Presentation The vice president of quality assurance presents, by exception, the status of the internal quality audits at the management review meetings.

Escalation A corrective action not completed according to plan is escalated to the president through the management review meetings.

Supplier Audits The vice president of quality assurance is also responsible for the management of key supplier audits. The key suppliers include assembly houses for outsourced builds, printed circuit board vendors, cable manufacturers, and memory SIMM module supplier. Supplier CARs (SCARs) are issued if required to the supplier for corrective and preventive action.

Critical Production Audit Areas A comprehensive set of procedures is in place to ensure that all customer and supplier material is properly safeguarded from electrostatic discharge throughout the manufacturing process flow. These include electrostatic awareness training, wrist, heel strap, and conductive floor wax monitoring, and grounding of floor and bench top mats.

A work instruction is used to measure and record solder paste deposition data for sample quantities of all jobs requiring the solder paste screen print process. A work instruction is also used to ensure that printed circuit boards meet or exceed customer or Growth cleanliness requirements. An omega meter is used to set up and audit the circuit board cleaning process on a daily basis.

Training All auditors receive internal quality audit training provided by the vice president of quality assurance. Additional external training is encouraged and sponsored by Growth.

8.2.3: Monitoring and Measurement of Processes

Audits As described in Section 8.2.2 of this manual, Growth performs an extensive first- and second-party audit program to make sure that Growth processes are efficient and effective. Growth also includes third-party and customer audit results into its strategic planning cycle.

Process Analytics Section 8.1 of this manual describes the way in which Growth uses analytical techniques to evaluate the effectiveness of its processes. The evaluations are based on an intensive set of quality objectives described in Section 5.4.1 of this manual.

Corrective Action As described in Section 8.5 of this manual, Growth employs an extensive system of corrective and preventive action, which includes a significant customer complaint response program, to investigate and resolve process-related issues.

8.2.4: Monitoring and Measurement of Product

Growth monitoring and measurement of product includes not only the incoming inspection and testing process discussed in Section 7.4 of this manual, but an extensive program of in-process and final inspection and testing of product. The vice president of quality assurance is responsible for all inspection and test procedures at Growth (refer to Inspection and Testing Processes).
🗁IandT.Doc

In Process Each key step of in-process inspection and testing is monitored by quality-control personnel. Specific check points include the following:

- *First article:* A first article inspection process performed on the first board of every production run. A distinct work tag is attached to identify that assembly. Authorization to commence the production run is given when the board passes this procedure.
- *Prereflow:* A prereflow solder inspection is performed on 100% of boards requiring the reflow soldering process. This inspection is performed immediately after the pick-and-place process by manufacturing personnel.
- *Component side:* A top-side in-process inspection is performed on 100% of boards built on either the prototype or production lines.
- *Prewave:* A prewave solder in-process inspection is performed on 100% of the boards requiring the wave solder process.

- *Bottom side:* When applicable, a bottom-side inspection is performed on boards built on either the prototype or production lines. This inspection is performed by quality-control personnel.
- *Special in-process tests:* Various types of electrical, functional, or visual testing is performed on a board at particular process locations as required. This testing either conforms to customer requirements or is deemed appropriate to ensure conformity to customer or Growth quality standards.

Final Inspection A final inspection is performed on 100% of boards built on either the prototype or production lines. All final inspections are performed by quality-control personnel. Finally assembled products are visually inspected by operators to verify that the products have the appropriate stamps or tags. Product that undergoes functional tests and burn-in are logged into the first pass yield database for analysis. All inspected and tested product receive a stamp to indicate completeness.

Records and Nonconformances Any material or assembly found in nonconformance is promptly identified and segregated as such. Prompt action is then taken to bring the material back into conformance or disposition according to the protocols for the control of nonconforming product as discussed in Section 8.3 of this manual.

The records of acceptance and/or rejection, which clearly show the responsible inspection authority, are maintained in their respective areas (i.e., either in materials or quality-assurance files).

Notification of products on hold is identified by a QC verbal communication to all affected areas. Any changes required to remove the product from hold is done through the ECO process.

All product that is shipped from Growth meets all required specifications. The shipper is required to verify the inclusion of all product and related components into the shipped package, including software and documentation as noted on the pick ticket for the order. The shipper validates the pick ticket with date shipped and his or her initials.

Software For required software, the software files to be replicated are revision controlled by means of the ECO process and its associated part number. Error checking is built into the replication process.

Records All records are maintained by either quality assurance or materials, as appropriate.

Customer Validation When required by contract, final acceptance occurs in the engineering validation phase, whereby the customer either attends the final testing or the testing is done at the customer's site, and the customer accepts or rejects the product's performance.

8.3: Control of Nonconforming Product

Process The process by which Growth identifies, documents, evaluates, separates, and disposes of nonconforming product, and notifies the appropriate executive functions, is contained within the document entitled "Control of Nonconforming Process." 🗁NCProcess.Doc

An MRB is used to make nonconforming product and material decisions. The group is represented by quality assurance, manufacturing, and continuing engineering, and the MRB report is used to notify the affected areas.

MRB Options Dispositions include rework, acceptance with or without repair by concession, and swap/replacement.

Concessions Concessions are achieved through direct contact between Growth's sales and marketing department and the customer.

Rework If a product fails in test and can be reworked, it is retested and reinspected. Products that cannot be reworked are tagged and segregated for further disposition by the MRB. For example, rejected products are identified and segregated from conforming product into one of three areas (i.e., the rework area on the manufacturing floor, the works in progress shortage rack on the manufacturing floor or the nonconformance storage cabinet in the stock room).

Failed Parts Other products, such as SIMM modules, cable assemblies, and components that fail in the process or in the field are dispositioned by segregating the material into an MRB location. Purchasing notifies the appropriate vendor for a return authorization number and returns the material to the vendor.

Records If agreed contractually, where nonconforming product is used as a result of concession with either the customer or the customer's representative, the nature of the nonconformity and the method of repair is recorded and reported to the customer. Sales and marketing handles this interface.

Reinspection In all cases, repaired and or reworked product is reinspected prior to shipment. When necessary, specific rework work instructions are used.

After-Sales Protocols Growth's after-sales process is described in Section 7.5.1 of this manual. After-sales activities are primarily the customer service manager's responsibility. Field data is key to the evaluation of product quality and reliability and is a bell weather for customer satisfaction or dissatisfaction.

8.4: Analysis of Data

Growth has described its extensive program to determine, collect, and analyze data, as a way to demonstrate the suitability and effectiveness of the GCQMS, in a number of sections in this manual. This information has been shown to be a key evaluation tool to establish where continual improvement of GCQMS effectiveness has occurred. In summary:

- Customer satisfaction is addressed in Section 8.2.1 of this manual.
- Conformity to product requirements is addressed in Section 7.2.1 of this manual.
- Trend analysis is addressed in Section 8.1 if this manual.
- Preventive action is address in Section 8.5.3 of this manual.
- The evaluation of suppliers is addressed in Section 7.4 of this manual.

In all cases, the evaluations are based on clearly described quality objectives, metrics, and targets, as addressed in Section 5.4.1 of this manual.

8.5: Improvement

8.5.1: Continual Improvement

Growth's vigorous and intensive program to continually improve GCQMS effectiveness is primarily based on the following inputs:

- Quality policy (see Section 5.3 of this manual);
- Quality objectives (see Section 5.4.1 of this manual);
- Audit results as a result of first-, second-, and third-party assessments (see Section 8.2.2 of this manual);
- Analysis of data (see Section 8.4 of this manual);
- Corrective and preventive actions (see Sections 8.5.2 and 8.5.3 of this manual);
- Management review by the GEMT (see Section 5.6 of this manual).

8.5.2: Corrective Action

Procedure The document describing corrective and preventive action is entitled "Corrective and Preventive Action Processes with Customer Complaints."
📖CAPA.Doc

This process of analysis includes the following programs:

- Audit program;
- Customer complaints;
- Customer returns;
- Product issue reports, which include software and hardware issues, and product feature requests;
- Customer/field reports;
- Engineering and documentation change control;
- First pass yield and quality data overview.

These areas form the data input for the corrective and preventative action program.

Responsibility The management of this program and the analysis of this data is the joint responsibility of the GEMT. The chairman of the group is the vice president of quality assurance.

Level of Action The level of corrective and/or preventive actions taken by Growth depends on its degree of impact on the product lines. These decisions are made by the GEMT. The GEMT can also assign decision making to local area managers, as required. Another way of resolving specific critical issues is by the formation of an engineering and support crash team when needed under the management of the vice president of design engineering.

Revisions The corrective and preventive action program uses the document and data control process to provide a means to change any documentation required as a result of this process.

In-Plant Operations In-plant operations include all in-plant inspection and acceptance testing, as well as the RMA system. All nonconformances are identified, corrected, and verified at Growth.

This data is contained in databases. These databases are managed by quality-assurance staff, who review and analyze the information. The vice president of quality-assurance reports these findings and the status of subsequent corrective action requests to top management in the management review meetings.

Handling CARs When a corrective action request is presented to a local area manager, it is the local area manager's responsibility to take timely action in defining and eliminating the root cause of the nonconformance.

Verification The presenter of the CAR is responsible for ensuring that the corrective action is taken and that it was effective, so that effective closure can occur.

Audit Program The vice president of quality assurance is responsible for managing the total quality audit program. This program includes: internal audits, SCARS, third-party audits, and customer audits. The status of this program is presented at the management review meetings.

Subcontractor Verification Corrective actions that result from subcontractor evaluation from results of incoming inspection, such as return to vendor (RTV) items, final test, or from on-site inspection, are managed by the vice president of quality assurance with the assistance of the purchasing manager.

Product Development Verification Corrective actions that result from engineering activities, which include the hardware and software databases, are the responsibility of the vice president of design engineering.

Managers All managers are responsible for the detection, analysis, and the eventual elimination of potential causes of nonconformities through the examination of available data. This data includes returned material, customer complaints, discrepant material reports, design review, and current documentation.

Plan All managers are responsible for generating the plan to remove potential causes of the nonconformity and for ensuring the plan results in the effective control of such actions.

Action All managers are responsible for collection and analysis of data within their respective areas. From the analysis of this data, the managers are responsible for deciding the appropriate action to be taken. These actions can include formation of quality improvement teams, assignment of tasks, and issuance of ECOS.

Customer Complaints

Responsibility: The customer service manager is responsible for collecting, analyzing, and generating corrective actions related to customer complaints. The customer complaints are obtained through the following sources:

- Direct customer contact via phone, fax, or e-mail;
- Field sales representatives or distributors.

Resolution: The corrective and preventive action programs are used to resolve customer complaints. Resolution is generally through the customer service complaints logs and the SCAR/CAR databases. Any form of customer feedback that is received can result in a CAR. As a result, this policy is designed to service the needs of the customer, yet minimize the burden placed upon the customer due to procedural requirements.

Status: The vice president of sales and marketing presents the status of customer complaints at the management review meetings.

8.5.3: Preventive Action

In addition to the general statements in regard to corrective and preventive action addressed in Section 8.5.2 of this manual, we note a few more specific activities related to preventive action at Growth.

Reporting The vice president of quality assurance coordinates the reporting at the management review on the status of any action plans that are taken in the area of preventive action.

Data Analysis For this purpose, data is analyzed from manufacturing, sales and marketing, quality assurance, and design engineering to detect and eliminate potential causes of nonconformities. Such data is displayed as Pareto charts on a monthly basis and reported as part of the management review process. A list of preventive actions is maintained to indicate progress in this important area.

Preventive actions are formed from all administrative and operational areas (e.g., improvements in the use of people, machines, instrumentation, facilities, and test procedures).

As with corrective actions, the scope of consideration, cost, and time spent in the resolution of preventive-action-related issues are proportional to the impact of such issues on the economic status of Growth. We do not wish to spend a million dollars to solve a one dollar problem. The GEMT is charged with this part of decision making. Records are maintained by the vice president of quality assurance for all corrective and preventive action activities.

CHAPTER

23

Contents

Case Study #2: Mike's Advice on ISO 9001:2000 from the Ground Floor Up

23.1 The Phone Call

Mike was quite satisfied with himself, not only because of his promotion to vice president of quality assurance—that was in the near past—but more so because he had led the activities to upgrade Growth to ISO 9001:2000 in such a way that company costs had been minimized. He pondered on whether hindsight would indicate how the original certification to ISO 9001:1994 could have been done in a less expensive manner. However, there was a lot of work to do, and he put the idea on the back burner for the present. Then, as Mike was sipping his morning decaf, black (a true macho man), the phone rang. He let it ring a few times while he got up the strength to talk, and then he answered in his usually highly professional manner, "Good morning, this is quality assurance at Growth, how may I help you?" The answer came back with a start, "Mike, it's Paul, you old smoothie, that's some phone technique. I'm impressed."

Mike and Paul had been university school chums when they majored in electrical engineering a good many years ago. Mike had gone on to earn a master's degree in industrial engineering, and Paul had gone on to earn an MBA. They hadn't seen each other in over 20 years.

"Paul, what in the world have you been up to? It's been a long, long time."

"Mike, we've got to get together some night to catch up over a couple of beers, but right now I'm in deep yogurt and maybe you can help me."

"Hey, for you old buddy, any time. But it has to be technical—no fixing you up and all that, like I used to."

"No, it's not that kind of problem. Besides you weren't all that great in the date department. Anyway, I'm happily married with three kids, and to make a long story short, I'm the president of a small design and manufacturing company called FastBoard, Incorporated. We're only 4 years old, but we have a super fast display system and some really hot customers. But they're small, so we started to bid the big OEMs. Next thing we know, we get an RFP that says you won't be considered unless you have an ISO 9001 certification. We can't believe it. We've got the best design at the lowest cost, and we're cut out. I need certification fast. Last week I read in the local newspaper about Growth's upgrade to ISO 9001:2000 and your 3 years of certification, so I'm hoping you can tell me your secret."

"Well, it always pays to have a consultant that you have a lot of confidence in, and respect, and we've been very lucky in that regard. His name is Sam, and I'll have him call you as soon as I get your phone number. He's been a great help."

"Mike, that's great. I've been searching for consultants, but I didn't really know any. What else do I need to do?"

"Paul, I'll tell you what. I've been thinking back on how we did so well in both certifications, and your phone call motivates me to put it down on paper. Perhaps it will help you, and I might submit it as a paper to our local quality group. It might be of interest to them, too. I'll put it together this weekend and get it off to you. I'm sure Sam will fill in the details when you two get together."

"Mike, I really appreciate this. The RFQ is due in 40 days, and maybe if I show them that FastBoard means business about getting certified, they'll waiver their Ts and Cs and let us bid. It's certainly worth a shot."

Mike wrote down Paul's phone number and address and called Sam. Fortunately, Sam had some consulting days open, and, after receiving Mike's suggestions, he and Paul met the following week to plan the program's details based on Mike's summary of necessary activities.

23.2 The Certification Plan from the Ground Floor Up

Mike did as he promised and sat down at his desk that Saturday morning and began to write the hindsight approach to ISO 9001:2000 that he would deliver to Paul over those few beers. He would fax the plan to Paul and Sam in advance for their review and concepts.

The plan would be somewhat difficult to write because he would have to integrate the ground-floor approach that Growth had taken for the ISO 9001:1994 version with the upgrade approach necessary for the ISO 9001:2000 version. As a result, he pulled out the ISO 9001:1994 records—about a three-foot-high pile of files—arranged the upgrade files along side the others and started to type into his computer. He decided to use a flow diagram for clarity. His work is shown in the flow chart labeled Figure 23.1.

Figure 23.1 Mike's ground-floor plan to achieve ISO 9001:2000 for FastBoard.

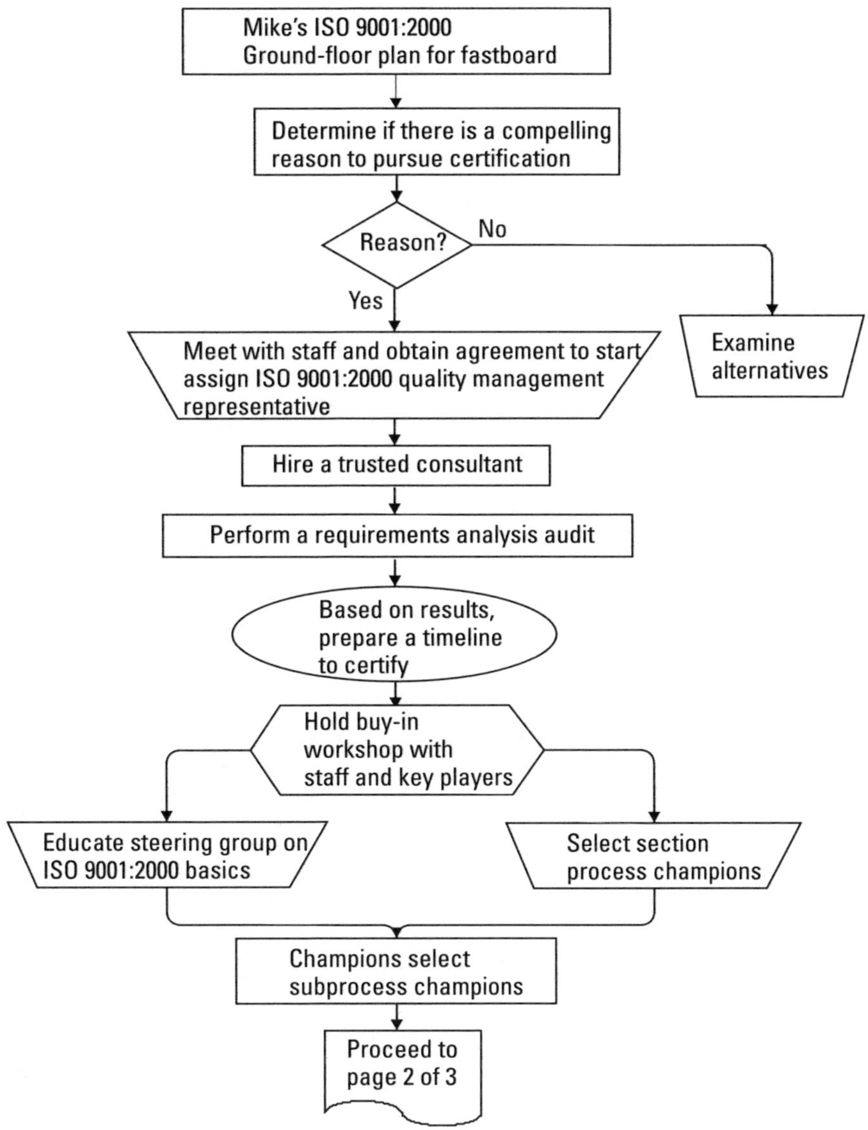

Figure 23.1 (continued)

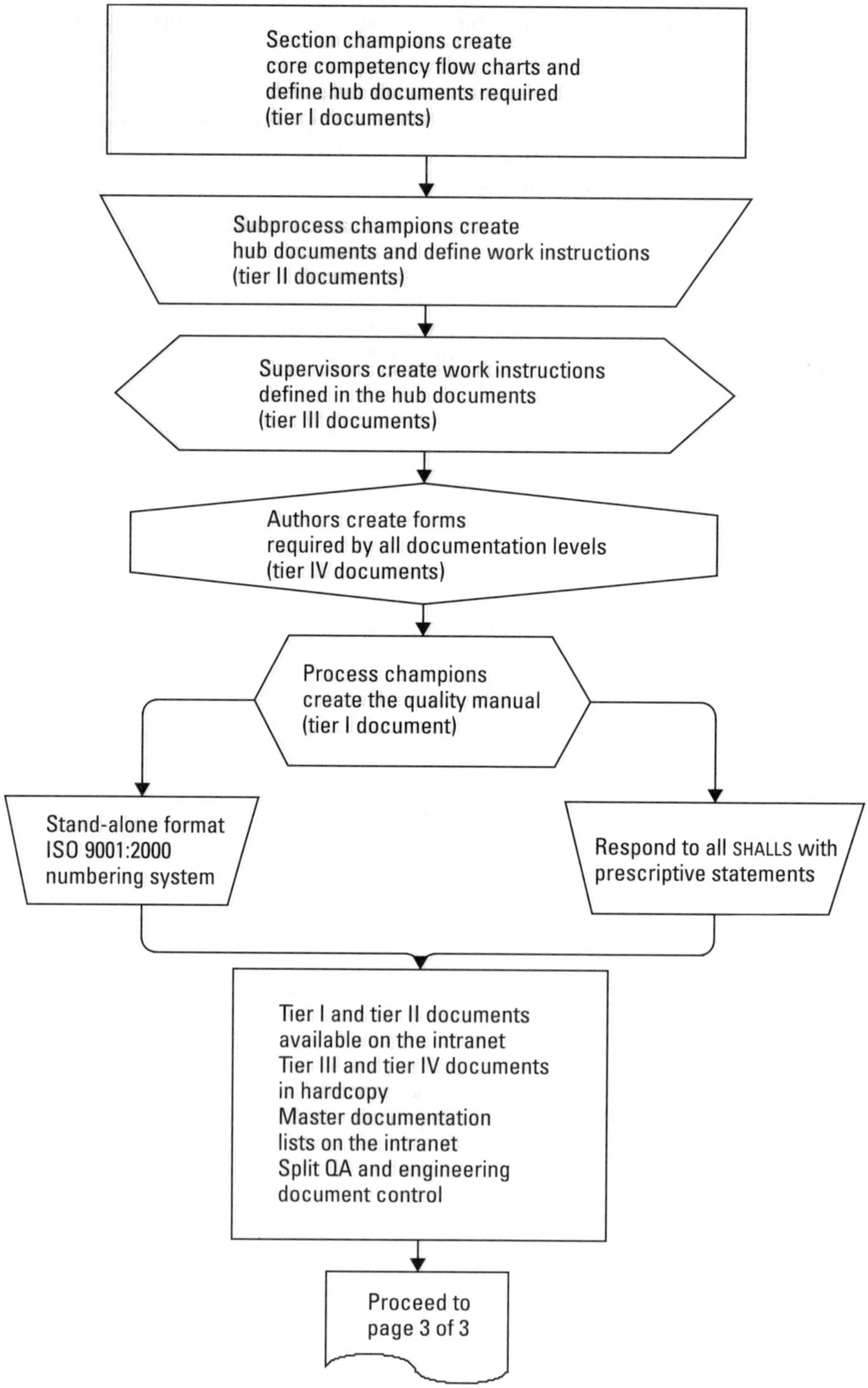

Figure 23.1 (continued)

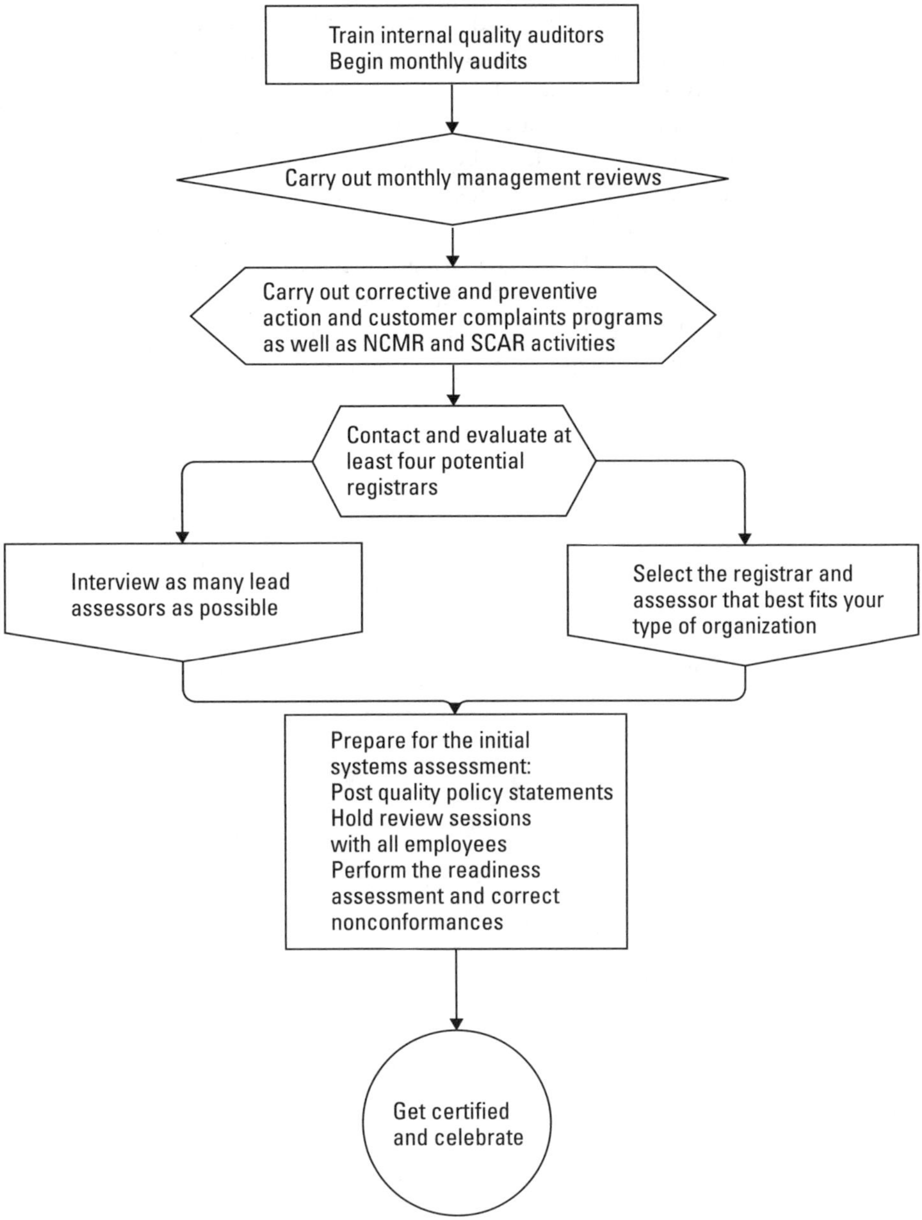

The Results Paul did receive a waiver from the OEM and won one of three prototype contracts as part of a runoff for the production contract. The ambitious certification program that Paul and Sam put together based on Mike's flow chart led to an ISO 9001:2000 certificate two months before the production RFQ was received. This was a little close, but it was in plenty of time for

FastBoard to submit and win 80% of the manufacturing award. The win established FastBoard as a player with the larger OEMs.

When Mike and Paul did get together, it turned out that Paul had married a woman that had been in his MBA class. Mike had married his high school sweetheart, had four kids, and his wife was a fashion artist that Paul's wife thought was the greatest pen-and-ink artist in the country.

Certification for FastBoard took 9 months. The major difference between Growth and FastBoard was that FastBoard celebrated with wine and cheese. As a matter of fact, FastBoard was different in another way. They didn't stuff the place with huge cookies like Growth; they used tons of cola. Either way, they were really high-energy people.

Appendix A: ISO 9000 Stewardship and Team Leader Summary

Ref. #:					
Date:	Supplier/Location:			Type: ISO 9001:2000	
Section ISO 9001	Subsection 9001	Clause 9001	Assignment of Stewards and Team Leaders to These Activities	Stewards by Name(s)	Team Leaders by Name(s)
4.0			Quality Management System		
	4.1		General Requirements		
	4.2		Documentation Requirements		
		4.2.1	General		
		4.2.2	Quality Manual		
		4.2.3	Control of Documents		
		4.2.4	Control of Records		
5.0			Management Responsibility		
	5.1		Management Commitment		
	5.2		Customer Focus		
	5.3		Quality Policy		
	5.4		Planning		
		5.4.1	Quality Objectives		
		5.4.2	QMS Planning		
	5.5		Responsibility, Authority, and Communications		
		5.5.1	Responsibility and Authority		
		5.5.2	Management Representative		
		5.5.3	Internal Communication		
6.0			Resource Management		
	6.1		Provision of Resources		
	6.2		Human Resources		
	6.3		Infrastructure		
	6.4		Work Environment		
7.0			Product Realization		
	7.1		Planning of Product Realization		

Ref #:
Date: **Supplier/Location:** **Type: ISO 9001:2000**

Section ISO 9001	Subsection 9001	Clause 9001	Assignment of Stewards and Team Leaders to These Activities	Stewards by Name(s)	Team Leaders by Name(s)
	7.2		Customer-Related Processes		
	7.3		Design and Development		
	7.4		Purchasing		
	7.5		Production and Service Provision (P&SP)		
		7.5.1	Control of P&SP		
		7.5.2	Validation of Processes for P&SP		
		7.5.3	Identification and Traceability		
		7.5.4	Customer Property		
		7.5.5	Preservation of Product		
	7.6		Control of Monitoring and Measuring (M&M) Devices		
8.0			Measurement, Analysis, and Improvement		
	8.1		General		
	8.2		M&M		
		8.2.1	Customer Satisfaction		
		8.2.2	Internal Audit		
		8.2.3	M&M of Processes		
		8.2.4	M&M of Product		
	8.3		Control of Nonconforming Product		
	8.4		Analysis of Data		
	8.5		Improvement		
		8.5.1	Continual Improvement		
		8.5.2	Corrective Action		
		8.5.3	Preventive Action		

Stewards Manage channels of information (i.e., policy, process, procedure, and forms) to ensure that the channel is fully documented, implemented, and demonstrating effectiveness in meeting the organization's quality objectives.

Team Leaders Formulate, implement, and effectively manage their specific projects (e.g., the quality manual, documentation control, records control, corrective and preventive action with customer complaints, and internal auditing).

Approved by: Distribution:

Dated:

Appendix B: Further Examples of Quality Policy Statements

ISO 9001:2000 Clause	ISO 9001:2000 Requirement	Quality Policy Statement
7.2.2: Partial Clause—Amendment to Product Requirements	In the case where product requirements are changed, we must ensure that both relevant documents are amended and relevant personnel are informed changes	Excellent's customer service representatives ensure that amendments to contracts are in written form by means of sales orders, reviewed by all affected departments by means of the request for quote (RFQ) report, and then accepted or rejected in written form and returned to the customer.
7.3.6: Partial Clause—Design and Development Validation	Product validation is to be performed in accordance with specifications to make sure that the product meets customer requirements	After acceptance by the chief engineer of the final design review and verification testing report, the project engineer contacts the customer and schedules an in-plant validation test with the customer present. The elevated temperature electrical and mechanical validation test results are compared to the customer's specifications, and upon acceptance the customer signs off on the final product release form.
4.2(c): Partial Clause—Control of Documents	Make sure that changes and current revision levels are identified	Document changes are accomplished using the document change form. Changes are reviewed by qualified personnel prior to approval. Revisions made to controlled documents are highlighted to the changes made. Revisions are tracked by the revision number and effective date. Changes to documentation are approved by those identified by management as the appropriate authority. The document control administrator maintains the authorization lists. Management makes the decision whether or not a change impacts the regulatory status of licensed product.
6.3: Partial Clause—Infrastructure	Determine, provide, and maintain the infrastructure needed to achieve conformance to product requirements	Production equipment and machines are regularly maintained by the maintenance department following the schedules and recommendations provided by their manufacturers or, if unavailable, the recommendations of an appropriate equipment engineer. Performance and accuracy of the equipment is continuously monitored by the production foremen.
8.2: Partial Clause—Monitoring and Measurement of Product	Records need to indicate the person(s) authorizing product release	100% inspection is performed on all boards built on either the prototype or production lines. The inspections are performed by quality control personnel. Pass/fail data is recorded on the traveler by means of QC stamps, and the records are maintained in the quality assurance laboratory files.

ISO 9001:2000 Clause	ISO 9001:2000 Requirement	Quality Policy Statement
6.2.2(b): Partial Clause—Competence, Awareness, and Training	Provide training or take other actions to fulfill training needs	The personnel department or representative at each of Excellent's plants is responsible for coordinating all training programs and for maintaining all records pertaining to training. This process is documented in SOP TR6.2-01. New employees at each plant receive, as a minimum, instruction on plant safety, the ISO 9000 quality system overview, skills instructions, and basic statistical concepts. The personnel department or representative at each plant surveys each department at least annually to identify on-the-job training and cross-training needs. The information is reported to the general manager who prepares the training plans and issues the training schedules to each department manager. Each department manager creates the required on-the-job training exercise to fulfill those needs and assigns an appropriately skilled trainer or trainers.
8.1: Partial Clause—General	Determine the appropriate analytical methods, including statistical techniques and the extent of their use	At the Excellent Corporation, the need for statistical or data analysis is decided on jointly by the quality assurance, engineering, and production departments in accordance with document QAP-8.1-01. The application of any statistical technique is directed for use only when it is clearly determined that such techniques will result in an improved process. Statistical and data analysis is required in the areas of process capability, machine control charts, in-process data collection records, quality reject records, inspection records, and material review board (MRB) records to ensure process reliability. Pareto and histograms are presented to management by the quality assurance department as part of the monthly review.

Appendix C: Checklist for ISO 9001:2000 Element 4.2.3: Control of Documents Quality Manual Requirements

Ref #:					
Date(s):	**Supplier Name/Location:**		**Type: ISO 9001:2000**		
Item	**Mandatory Activities To Be Covered in the Quality Manual**	**First Draft**	**First Edit**	**Final Edit**	**Release Date Plan/Actual**
1.0	Define the tier II process documents (SOP procedure)				
1.1	Controlled document attributes				
	Quality manual				
	Tier II processes or procedures				
	Tier III work instructions				
	Tier IV forms				
1.2	Hard-copy versus electronic media				
	Access limits				
	Backup protocols				
	Read/write protocols				
1.3	How documents are created				
1.4	Review and approval protocols				
	QA documents				
	Engineering documents				
	Documents of external origin				
	Reapproval protocols				

Ref #: Date(s):	Supplier Name/Location:		Type: ISO 9001:2000		
Item	**Mandatory Activities To Be Covered in the Quality Manual**	**First Draft**	**First Edit**	**Final Edit**	**Release Date Plan/Actual**
1.5	Mix (manuals versus individual documentation)				
	Tier II master list				
	Tier III master list				
	Tier IV master list				
1.6	Method of distribution/removal				
1.7	Control method (central versus local area manager)				
	QA documents				
	Engineering documents				
	Documents of external origin				
1.8	Obsolete document protocols				
1.8.1	Normal obsolete removal				
1.8.2	Retained for info/legal purposes				
1.9	Revision protocols				
1.9.1	Review and approval				
1.9.2	Nature of change control				
1.9.3	Pertinent background protocol				
2.0	Method of records control				

Appendix D: An Example of Excellent's Process Flow-Charting Protocol

Hub Document for Quality Audits

Quality Audit Processes

Text Supplement for the Quality Audit Flow Charts

The flow chart entitled "Internal and External Audit Process, File XAUDIT. ACL" describes the quality-assurance audit process in detail and introduces the safety audit process, which is found in the flow chart entitled "Safety Audit Process, File SAFETY.ACL." In all cases, the flow chart takes precedence over this text (See Figure D.1).

The senior lead auditor manages both programs.

Lead Auditor Training

Each lead auditor completes the training program shown in the workbook entitled "Systems Lead Auditor Training Manual." This program may be run by either an outside qualified source or by the senior lead auditor.

An outside qualified source must either be individually certified under a national or international certification schema or be an employee of an accredited training program (e.g., a RAB-certified lead assessor, or RAB-accredited training program).

Abbreviations in figure

QA = quality assurance
CARs = corrective action reports
SCAR = supplier corrective action report
PA = purchasing agent
SC = steering committee

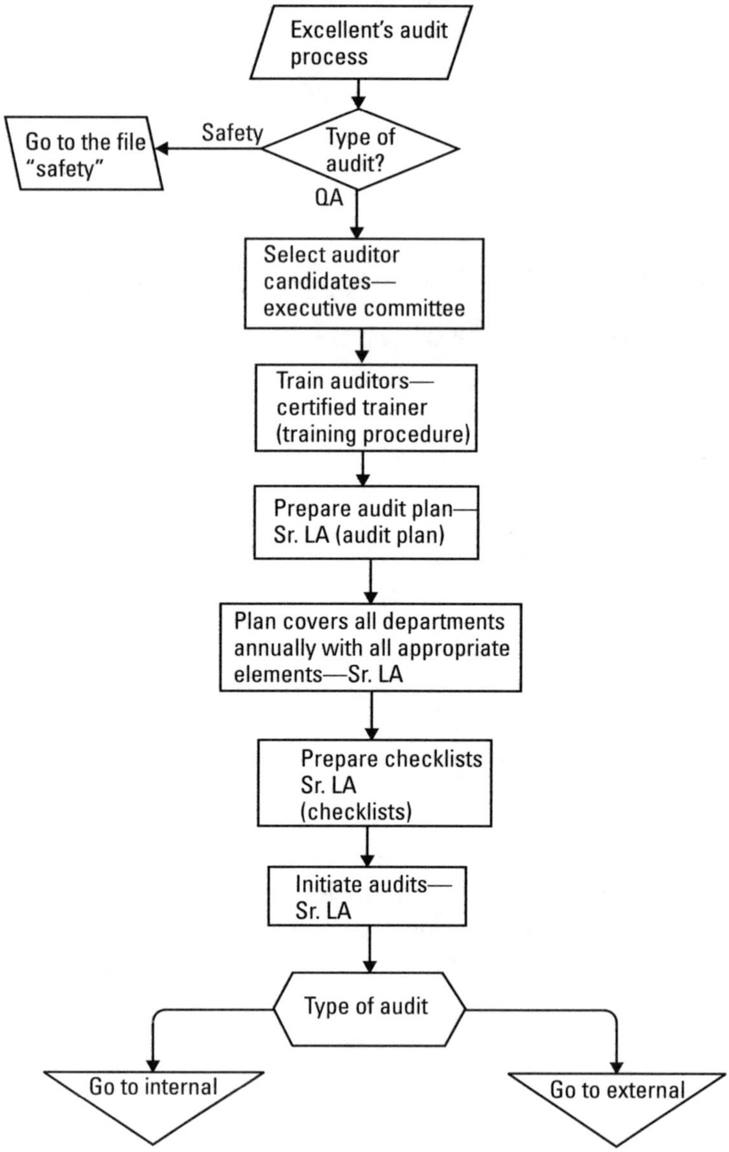

Figure D.1 Excellent's quality audit process flow chart.

About the Author

ISO 9000 Lead Auditor

Jay J. Schlickman, B.S., M.S., CQMS-LA, is a RAB-certified quality management systems lead auditor. He has completed 400 third-party ISO 9000 assessments and served as lead assessor/team leader/supervisor on 376 of those assessments. He also served as an A.I.A.G.-qualified QS-9000 automotive assessor (April 1995–December 1999). He has supported the accredited certification of 106 companies to date (sum of third-party assessing and consulting). He is a RAB-qualified ISO 9001:2000 auditor and a certified auditor under the Canadian Medical Devices Conformity Assessment System (CMDCAS). This includes 10 ISO 9001:2000 certifications (sum of consulting, initial systems assessments, and upgrades).

Quality Management System Consulting

His quality-management background includes 35 years in the combined implementation and facilitation of MIL-Q-9858A, FDA/CGMP 820, EN46001, ISO 13485/8, MDD 93/42/EEC, MIL-STD-1772, ISO 9001, QS-9000, ISO 14001, AS9000, and TQM quality systems. As a result of this extensive quality background, he has also provided significant consulting support to companies either in the process of achieving ISO 9000/QS-9000 certification, having achieved certification, or in postcertification process development—34 companies and 45 sites to date.

Industrial Experience

Due to a very wide range of high-tech management experience covering over 40 years, his SIC code coverage includes aerospace, computer software, electrical and electromechanical products, electronic components and products, including computers, health/medical electronic products, and optics, and test laboratories. Other areas include the paper, laboratory-purpose-bred animal, and construction industries.

General Management

Mr. Schlickman's high-tech management responsibilities have ranged from design engineer to president and CEO in multiplant and multidivisional organizations, managing groups as large as 300 employees.

Education and Scientific Awards

Mr. Schlickman received B.S. and M.S. degrees in physics from Northeastern University, Boston, Massachusetts. He has served as an infrared detector NASA consultant, presented and published papers in the fields of solid-state physics, quantum electronics, and quality management systems, and received a Honeywell Engineering and Scientist award for outstanding achievements in the computer-aided design of infrared detectors used in industrial and military applications. He is the author of *ISO 9000 Quality Management System Design*, and has been jointly awarded Multispectral Detector U.S. Patent #3,761,718.

Index

Recent Titles in the Artech House Technology Management and Professional Development Library

Bruce Elbert, Series Editor

Critical Chain Project Management, Lawrence P. Leach

Decision Making for Technology Executives: Using Multiple Perspectives to Improve Performance, Harold A. Linstone

Designing the Networked Enterprise, Igor Hawryszkiewycz

Engineering and Technology Management Tools and Applications, B. S. Dhillon

The Entrepreneurial Engineer: Starting Your Own High-Tech Company, R. Wayne Fields

Evaluation of R&D Processes: Effectiveness Through Measurements, Lynn W. Ellis

From Engineer to Manager: Mastering the Transition, B. Michael Aucoin

Introduction to Information-Based High-Tech Services, Eric Viardot

Introduction to Innovation and Technology Transfer, Ian Cooke and Paul Mayes

ISO 9001:2000 Quality Management System Design, Jay Schlickman

Managing Complex Technical Projects: A Systems Engineering Approach, R. Ian Faulconbridge and Michael J. Ryan

Managing Engineers and Technical Employees: How to Attract, Motivate, and Retain Excellent People, Douglas M. Soat

Managing Successful High-Tech Product Introduction, Brian P. Senese

Managing Virtual Teams: Practical Techniques for High-Technology Project Managers, Martha Haywood

Mastering Technical Sales: The Sales Engineer's Handbook, John Care and Aron Bohlig

The New High-Tech Manager: Six Rules for Success in Changing Times, Kenneth Durham and Bruce Kennedy

Planning and Design for High-Tech Web-Based Training, David E. Stone and Constance L. Koskinen

Preparing and Delivering Effective Technical Presentations, Second Edition, David Adamy

Reengineering Yourself and Your Company: From Engineer to Manager to Leader, Howard Eisner

Successful Marketing Strategy for High-Tech Firms, Second Edition, Eric Viardot

Successful Proposal Strategies for Small Businesses: Using Knowledge Management to Win Government, Private Sector, and International Contracts, Third Edition, Robert S. Frey

Systems Engineering Principles and Practice, H. Robert Westerman

Team Development for High-Tech Project Managers, James Williams

For further information on these and other Artech House titles, including previously considered out-of-print books now available through our In-Print-Forever® (IPF®) program, contact:

Artech House
685 Canton Street
Norwood, MA 02062
Phone: 781-769-9750
Fax: 781-769-6334
e-mail: artech@artechhouse.com

Artech House
46 Gillingham Street
London SW1V 1AH UK
Phone: +44 (0)20 7596-8750
Fax: +44 (0)20 7630-0166
e-mail: artech-uk@artechhouse.com

Find us on the World Wide Web at:
www.artechhouse.com